Norton Twins Owners Workshop Manual

by D J Rabone B. Tech.

Models covered:

497 cc	Model 88	1957-1963
597 cc	Model 99	1957-1966
*597 cc	Model 77	1957-1959
497 cc	Model 88SS	1961-1966
597 cc	Model 99SS	1962-1966
647 cc	Sports Special	1962-1968
647 cc	Manxman	1962-1968
745 cc	Atlas	1963-1968
*745 cc	N.15	1967-1967
*745 cc	G.15 (Matchless)	1964-1967
*745 cc	P.11	1966-1967
647 cc	Mercury	1968-1970
*745 cc	Ranger (P11A)	1967-1968

Not F/Bed

ISBN 978 0 85696 187 8

© Haynes Group Limited 1990

All rights reserved. No part of this book may be reproduced or transmitted in
any form or by any means, electronic or mechanical, including photocopying,
recording or by any information storage or retrieval system, without permission
in writing from the copyright holder.

(187-12AP7)

THE
BOOK

Haynes Group Limited
Haynes North America, Inc

www.haynes.com

Acknowledgements

Our grateful thanks are due to Norton-Villiers Limited for technical assistance given and permission to use many of their illustrations. Brian Horsfall gave necessary assistance with the overhaul and devised ingenious methods for overcoming the lack of service tools. Les Brazier took the photographs that accompany the text. Jeff Clew edited the text after it had been checked by Les Emery of Fair Spares, Rugeley, Staffs, incorporating his many useful suggestions.

Our thanks are also due to Tim Marsh who kindly loaned us the 750cc Atlas model on which this manual is based and to Bob Stapleford, Norton enthusiast, for the loan of his files on Norton Twins. The 750cc Atlas featured on the front cover was supplied by Graham Garrett of West Camel.

Finally, we would also like to acknowledge the help of the Avon Rubber Company who kindly supplied illustrations and advice about tyre fitting and of Amal Limited for the use of their carburettor illustrations.

About this manual

Although this manual is based on the stripdown and reassembly of a Norton "Atlas" twin, covering every major assembly, the construction of all the heavyweight twins is so similar that few of the procedures outlined cannot be applied. Where differences occur, separate notes are made in the text. The more recent "Commando" models are not covered, however, because they form the subject of another Haynes Owners Workshop Manual in this series.

Where only a partial stripdown is required, it is recommended that the procedure for the stripdown and subsequent reassembly is read first, so that it can be determined which parts of the machine will have to be removed in order to gain access. For example, if a magneto/distributor overhaul is necessary, there is no requirement to remove the engine from the frame, although it will become apparent that it is necessary first to remove the timing cover, automatic timing unit and in some cases the carburettors (to provide access) as a preliminary.

A 650cc Norton was the proud possession of the author for two years and was one of the machines on which his motor-cycle mechanics was learnt - by trial and error, an expensive and time consuming method! All the tips and warnings gained during the ownership of this machine are included in the text and with the help of this manual and a little patience, the owner should find maintaining his own Norton Twin an interesting hobby.

Unless specially mentioned and therefore considered essential, Norton service tools have not been used. There are invariably alternative means of loosening or slackening some vital component, when service tools are not available and risk of damage is to be avoided at all costs.

Each of the six chapters is divided into numbered sections. Within the sections are numbered paragraphs. Cross-reference throughout this manual is quite straightforward and logical. When reference is made, 'See section 6.10'.it means section 6, paragraph 10 in the same chapter. If another chapter were meant it would say, 'See Chapter 2, section 6.10'.

All photographs are captioned with a section/paragraph number to which they refer, and are always relevant to the chapter text adjacent.

Figure numbers (usually line illustrations) appear in numerical order, within a given chapter. 'Fig 1.1' therefore refers to the first figure in Chapter 1.

Motorcycle manufacturers continually make changes to specifications and recommendations, and these, when notified, are incorporated at the earliest opportunity.

Left hand and right hand descriptions of the machines and their components refer to the left and right of a given machine when the rider is seated normally.

Whilst every care is taken to ensure that the information in this manual is correct no liability can be accepted by the author or publishers for loss, damage or injury, caused by any errors in or omissions from the information given.

Contents

Norton Atlas 750 cc

Introduction to the Norton Twins

The series of large capacity twin cylinder machines which culminated in the Norton Atlas, and its derivatives brought about by the marrying of a production engine with a frame developed from Nortons' legendary Manx racers. The engine, a 497cc parallel twin, had been developed by Nortons in the immediate post war years to counter the success of another British manufacturer who had introduced a highly attractive design in 1937. This first Norton twin was unveiled at the 1949 Earls' Court Show as the Model 7.

Its specification included a cast iron engine with splayed exhaust ports and pushrods located in cast-in tunnels at the front of the cylinder block. The magneto and camshaft were driven by separate chains that formed part of the timing gear assembly and were fully enclosed within a conventional timing cover.

The famous Norton Roadholder forks looked after the front suspension and were attached to what was colloqually known as a 'garden gate' trame employing rear suspension units of the plunger type, typical of that era.

Shortly afterwards, the Norton twins assumed the name "Dominator" and were fitted with an entirely new type of frame having rear suspension of the swinging arm type. Because this frame gave a much more comfortable ride and improved road holding, it became known as the 'featherbed' frame. It is alleged this description was originally coined by the late Harold Daniell when he tested one of the new frames fitted on that occasion to a 'Manx' racer. Alloy engine components were introduced around the same time and in 1951, the model 88 appeared, initially destined for export only. It was virtually a model 7 engine and gearbox fitted into a 'featherbed' frame, with shortened 'Roadholder' forks - a high speed road burner with handling to match!

Four years later full width hub brakes were included in the specification. This produced a machine which has become a standard against which all other motor cycles are measured and a special builders' dream. No higher compliment could a manufacturer achieve. The "Unapproachable Norton" continued in both production and racing guises.

Further development of the Model 88 led to a "big brother" in 1956 when the Model 99 became available. With a 597 cc engine, it was a large capacity machine for its day and it became a favourite with sidecar owners. For the latter, it was produced without the "Featherbed" frame as the model 77. Also, in 1958, Nortons began replacing the magneto/dynamo electrical system with either hybrid magneto /AC generator systems or A.C. generator/coil ignition systems.

In 1960, the frame underwent its most major modification when the top tubes were brought closer together, allowing a narrower tank. This modification become known as the Slimline Featherbed'. Enclosed panelling was used briefly at the rear end of the machines.

It was now time for the engine to show that it deserved the frame and forks that it had inherited from its famous stable mates. A much modified '88' was taken to the Isle of Man and entered in the 1961 TT where it became the first pushrod engined machine to lap the Island at over 100 mph. It came third in the senior race ridden by Tom Phillis. Developments from this machine were incorporated in the production racers sold by Nortons' and in 1962, the down-draught inlet manifold head, which produces so much more power from these engines, became available on the 'SS' versions of the Model 88 and 99. In addition a 646 cc Dominator was announced and followed shortly after by the last of this line, the Atlas, with a 745cc version of the Dominator engine.

Financial problems had necessitated Norton Motors becoming part of the Associated Motor Cylcles Group and in 1962 the old Bracebridge Street works in Birmingham, the traditional home of the Norton, finally closed when all production was moved to Woolwich. Although the Norton range of motor cycles was continued, some hybrids were marketed comprising in several instances Norton engines fitted into AMC cycle parts. The twins were, however, the least affected, apart from some export only versions, such as the Matchless framed Norton N15 and the Matchless G 15, both of which utilised the Atlas engine unit. In addition, the Norton P11A, combining the Atlas unit in a scrambles frame, completed the hybrids.

During the middle sixties, competition from the Japanese motor cycle industry became intense and Associated Motor Cycles, with its obsolete range of small bikes and its limited production of large bikes in the pre-'Superbike' era, was faced with severe financial pboblems such that in 1966, the company called in the receiver. However, towards the end of 1966 a satisfactory arrangement was completed for their assets to be acquired by Manganese Bronze Holdings Limited, of which Villiers Engineering was a subsidiary. A new company was born - Norton-Villiers Ltd and the Dominator and its derivatives continued to flow until the early 1970s, with models such as the 650 cc Mercury the 750 cc Ranger for export and the Atlas as the mainstay of the home market.

From the owners viewpoint, the Dominators are not the easiest engines to work on. The integrated cylinder head and rocker box and the arrangement of the cylinder head studs make this one of the most difficult heads to remove, especially in the case of the larger capacity engines. However, once the head is off (on the larger machines) the engine is easily removed as a unit with the gearbox - or by itself - and is soon stripped. Also, the frame and forks offer no mechanical problems.

Once on the move, these long wheelbase machines can be a problem to handle in towns and traffic especially if fitted with clip-on handlebars; but it must be remembered, this is a race bred motor cycle which loves the open road and was conceived before the 70 mph speed limit was introduced. It corners at over 100 mph-in the words of every Norton owner -"like it was on rails". An overall comment about these machines would be that they are sporting mounts whose reliability and handling have been proved without doubt by the scores of races they have won at all levels of production racing and the success of the "Domi-Racers" of the mid'60s. If maintained as described in this manual, with a little patience and care, the owner will have an excellent, reliable, fast and safe machine.

Modifications to the Norton Twins

The Dominator twins have been manufactured for over 24 years. During this time, modifications have taken place and only a brief account of these changes can be included here.

Frame:
The original Model 7 was introduced in 1949 and featured plunger rear suspension, garden gate frame and 'Roadholder' forks. In 1951, the engine was also offered mounted in a development of the Manx frame known as the Featherbed. This featured a pivoted rear fork and shortened "Roadholder" forks. The pivoted rear fork was introduced on the "garden gate" Nortons in 1953 and, although the Model 7 (or Domi 7) was discontinued in 1956, the frame continued in use powered by a '99' engine until 1959. The latter machine was known as the Model 77 and was primarily used for sidecar work. In 1960, the top frame tubes were brought closer together to allow a narrower tank and this development was known as the "Slimline Featherbed". All Dominators were housed in this frame until 1966 when the Atlas engine was slotted into a Matchless frame to produce the Norton N15 and Matchless G15 machines. In 1969, a lightweight trials frame was devised to produce the P11A which developed into the "Ranger". All these machines utilised versions of the "Roadholder" fork and Norton wheel hubs.

Electrics.
The original Dominators had a dynamo mounted forward of the cylinders to power the lights, and magneto ignition. In 1958, the dynamo was replaced by a crankshaft driven alternator and some models of the range had the magneto replaced by a coil ignition unit. The practice of using coil or magneto ignition varied from model to model, year to year, until 1967; by this time, all machines had coil ignition.

Carburettors:
All models have been fitted with 'Amal' instruments. The Amal Monobloc was used for the 10 years 1955-65. Previous to this, the carburettor design featured a single, separate float chamber unit. Twin carburettors became available on the 'SS' versions from 1962 onwards although many twin carburettor conversions were available for the standard head. The Monobloc carburettor required "handing" in order to produce a twin carburettor set-up - this was remedied in 1966 by the introduction of the Concentric carburettor which incorporates the float chamber below the main body.

Gearbox:
Prior to 1956, all machines used Burman gearbox units. However, following with the amalgamation with AMC Ltd a neater unit was produced which, in a developed form, was used in the later Commando. This unit retained the Norton clutch.

Wheels:
The only major modification that has occurred to the wheels was the introduction of full-width aluminium alloy hubs in 1956. Brakes have remained standard since this time although many firms, including the manufacturer, offer twin leading shoe front brake plates.

Engine:
The latest Dominator engines are very similar to the early models except that dimensions have increased as the engines have grown in capacity (see Specifications - Chap 1). The most major design changes have occurred to the cylinder head and camshaft. In 1955, the cast iron cylinder head was replaced by an aluminum alloy casting. This was developed such that all post 1960 cylinder heads give a higher compression ratio. In addition, after racing development, the 'SS' head with its downdraught inlet tracts and wider splayed exhaust ports was introduced in 1962. This modification, coupled with the sports camshaft which had been introduced in 1956, produced a large power increase from these engines.

General Notes:
1. Splayed exhaust ports produce better engine cooling and increased exhaust scavenging by reducing the bend in the exhaust pipe at the head.
2. The 1961 Dominator SS had small bore siamesed exhausts with a special silencer but did not necessarily have the true 'SS' head.
3. The 650cc engine incorporates a larger throw crank with short connecting rods and piston skirts. The cylinder barrel has bigger connecting rod clearance slots.

Ordering spare parts

When wishing to purchase spare parts for the Norton heavyweight twins it is best to deal direct with a Norton specialist. Parts cannot be obtained direct from Norton Motors Limited. When ordering parts, always quote the frame and engine numbers in full, without omitting any prefixes or suffixes. It is also advisable to include note of the colour scheme of the machine if the parts ordered are required to match.

The engine number is stamped on the left hand side of the crankcase immediately below the base of the cylinder barrel.

The frame number is stamped on the left hand, rear, gusset plate on featherbed frames or on the steering head of scrambles type frames.

Always fit parts of genuine Norton manufacture. Pattern parts may be available sometimes at lower cost, but they do not necessarily make a satisfactory replacement for the originals.

There are cases where reduced life or sudden failure has occurred, to the overall detriment of performance and perhaps rider safety.

Retain any worn or broken parts until the replacements have been obtained; they are sometimes needed as a pattern to help identify the correct replacement when design changes have been made during a production run.

Some of the more expendable parts such as spark plugs, bulbs, tyres, oils and greases etc., can be obtained from accessory shops and motor factors, who have convenient opening hours, charge lower prices and can often be found not far from home. It is also possible to obtain parts on a Mail Order basis from a number of specialists who advertise regularly in the motor cycle magazines.

Engine number location

Frame number location

Routine maintenance

Periodic routine maintenance is a continuous process that commences immediately the machine is used. It must be carried out at specified mileage recordings or on a calendar basis if the machine is not used frequently, whichever falls soonest. Maintenance should be regarded as an insurance policy, to help keep the machine in the peak of condition and to ensure long, trouble-free service. It has the additional benefit of giving early warning of any faults that may develop and will act as a regular safety check, to the obvious advantage of both rider and machine alike.

The various maintenance tasks are described under their respective mileage and calendar headings. Accompanying diagrams are provided, where necessary. It should be remembered that the interval between the various maintenance tasks serves only as a guide. As the machine gets older or is used under particularly adverse conditions, it would be advisable to reduce the period between each check.

Some of the tasks are described in detail, where they are not mentioned fully as a routine maintenance item in the text. If a specific item is mentioned but not described in detail, it will be covered fully in the appropriate Chapter. No special tools are required for the normal routine maintenance tasks. The tools contained in the kit supplied with every new machine will prove adequate for each task or if they are not available, the tools found in the average household.

Weekly or every 250 miles

Check oil tank level and replenish if necessary with engine oil of the recommended grade. Note: the level should be checked after the engine had run for at least a minute, especially if the machine has been idle for more than a week.

Grease brake pedal pivot and oil all other exposed control cables and joints.

Check the battery electrolyte level, chain adjustments and tyre pressures.

Check tension of primary and secondary chains.

Monthly or every 1000 miles

Complete the maintenance tasks listed under the preceding weekly heading, then the following additional items:

Check and, if necessary, adjust both brakes.

Check primary chaincase oil level and top up if necessary. Check nuts and bolts for tightness.

Three monthly or every 2500 miles.

Complete all the checks listed under the weekly and monthly headings, then the following items:

Drain the oil tank whilst the oil is warm and remove and clean the filters before refilling with new oil of the correct viscosity.

Check the gearbox oil level and top up if necessary.

Remove and clean both spark plugs and reset the gaps.

Check the ignition timing after adjusting the contact breaker gaps.

Remove and lubricate the final drive chain; change the primary chaincase oil. Check the clutch adjustment.

Six monthly or every 5000 miles

Again, complete all the maintenance tasks listed previously, they complete the following additional tasks:

Change the gearbox oil, and also the oil in the front forks.

Check and adjust the camshaft chain.

Clean the contact breaker points and lubricate the contact breaker cam and the auto-advance unit (where not fitted in timing chest).

Grease the brake operating arm pivots (grease sparingly to prevent grease reaching the brake linings).

Check and if necessary, adjust the valve clearances.

Fit a new air filter element.

RM1 Do not refill beyond top rib of oil tank

RM2 Refill or top-up gearbox through inspection cover

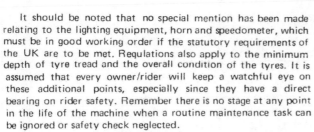

RM3 Chaincase level plug is below footrest. Note rubber plug for clutch adjustment

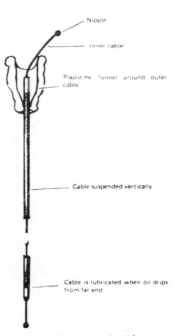

RM4 Oiling control cables

It should be noted that no special mention has been made relating to the lighting equipment, horn and speedometer, which must be in good working order if the statutory requirements of the UK are to be met. Regulations also apply to the minimum depth of tyre tread and the overall condition of the tyres. It is assumed that every owner/rider will keep a watchful eye on these additional points, especially since they have a direct bearing on rider safety. Remember there is no stage at any point in the life of the machine when a routine maintenance task can be ignored or safety check neglected.

carburettors. Check that they are synchronised correctly.

Check the swinging arm bushes for play and fill the pivot housing with grease. (Metal bushes only).

Grease the speedometer drive.

Yearly or every 10,000 miles

After completing the weekly, monthly, three-monthly and six-monthly tasks, continue with the following additional items: Remove and oil all the control cables. Remove and repack the wheel bearings with grease. Check the primary and secondary chains and sprockets for wear, also both carburettors. If performance has fallen off, decarbonise the engine and regrind the valves.

Lubrication Chart

WHEEL BEARINGS

TELESCOPIC FORKS

PRIMARY CHAINCASE. (L.H. SIDE)

OIL TANK

SWINGING ARM PIVOT

GEARBOX

SPEEDOMETER DRIVE

WHEEL BEARINGS

FINAL DRIVE CHAIN (L.H. SIDE)

ALSO: CONTROL CABLES, CENTRE AND PROP STAND PIVOTS AND CONTACT BREAKER CAM

Recommend Lubricants

Component	Grade	Quantity
Engine (oil tank)	Castrol GTX	4½ Imp pints
Gearbox	Castrol Hypoy	1 Imp pint
Front forks	Castrolite	150 cc per leg*
Primary chaincase	Castrolite	130 cc
Rear chain	Castrol Graphited Grease	
Wheel hubs, swinging arm, speedometer drive, contact breaker cam and other grease points	Castrol LM Grease	

*170 cc per leg, scrambles models

Safety first!

Professional motor mechanics are trained in safe working procedures. However enthusiastic you may be about getting on with the job in hand, do take the time to ensure that your safety is not put at risk. A moment's lack of attention can result in an accident, as can failure to observe certain elementary precautions.

There will always be new ways of having accidents, and the following points do not pretend to be a comprehensive list of all dangers; they are intended rather to make you aware of the risks and to encourage a safety-conscious approach to all work you carry out on your vehicle.

Essential DOs and DON'Ts

DON'T start the engine without first ascertaining that the transmission is in neutral.

DON'T suddenly remove the filler cap from a hot cooling system – cover it with a cloth and release the pressure gradually first, or you may get scalded by escaping coolant.

DON'T attempt to drain oil until you are sure it has cooled sufficiently to avoid scalding you.

DON'T grasp any part of the engine, exhaust or silencer without first ascertaining that it is sufficiently cool to avoid burning you.

DON'T allow brake fluid or antifreeze to contact the machine's paintwork or plastic components.

DON'T syphon toxic liquids such as fuel, brake fluid or antifreeze by mouth, or allow them to remain on your skin.

DON'T inhale dust – it may be injurious to health (see *Asbestos* heading).

DON'T allow any spilt oil or grease to remain on the floor – wipe it up straight away, before someone slips on it.

DON'T use ill-fitting spanners or other tools which may slip and cause injury.

DON'T attempt to lift a heavy component which may be beyond your capability – get assistance.

DON'T rush to finish a job, or take unverified short cuts.

DON'T allow children or animals in or around an unattended vehicle.

DON'T inflate a tyre to a pressure above the recommended maximum. Apart from overstressing the carcase and wheel rim, in extreme cases the tyre may blow off forcibly.

DO ensure that the machine is supported securely at all times. This is especially important when the machine is blocked up to aid wheel or fork removal.

DO take care when attempting to slacken a stubborn nut or bolt. It is generally better to pull on a spanner, rather than push, so that if slippage occurs you fall away from the machine rather than on to it.

DO wear eye protection when using power tools such as drill, sander, bench grinder etc.

DO use a barrier cream on your hands prior to undertaking dirty jobs – it will protect your skin from infection as well as making the dirt easier to remove afterwards; but make sure your hands aren't left slippery. Note that long-term contact with used engine oil can be a health hazard.

DO keep loose clothing (cuffs, tie etc) and long hair well out of the way of moving mechanical parts.

DO remove rings, wristwatch etc, before working on the vehicle – especially the electrical system.

DO keep your work area tidy – it is only too easy to fall over articles left lying around.

DO exercise caution when compressing springs for removal or installation. Ensure that the tension is applied and released in a controlled manner, using suitable tools which preclude the possibility of the spring escaping violently.

DO ensure that any lifting tackle used has a safe working load rating adequate for the job.

DO get someone to check periodically that all is well, when working alone on the vehicle.

DO carry out work in a logical sequence and check that everything is correctly assembled and tightened afterwards.

DO remember that your vehicle's safety affects that of yourself and others. If in doubt on any point, get specialist advice.

IF, in spite of following these precautions, you are unfortunate enough to injure yourself, seek medical attention as soon as possible.

Asbestos

Certain friction, insulating, sealing, and other products – such as brake linings, clutch linings, gaskets, etc – contain asbestos. *Extreme care must be taken to avoid inhalation of dust from such products since it is hazardous to health.* If in doubt, assume that they *do* contain asbestos.

Fire

Remember at all times that petrol (gasoline) is highly flammable. Never smoke, or have any kind of naked flame around, when working on the vehicle. But the risk does not end there – a spark caused by an electrical short-circuit, by two metal surfaces contacting each other, by careless use of tools, or even by static electricity built up in your body under certain conditions, can ignite petrol vapour, which in a confined space is highly explosive.

Always disconnect the battery earth (ground) terminal before working on any part of the fuel or electrical system, and never risk spilling fuel on to a hot engine or exhaust.

It is recommended that a fire extinguisher of a type suitable for fuel and electrical fires is kept handy in the garage or workplace at all times. Never try to extinguish a fuel or electrical fire with water.

Note: *Any reference to a 'torch' appearing in this manual should always be taken to mean a hand-held battery-operated electric lamp or flashlight. It does **not** mean a welding/gas torch or blowlamp.*

Fumes

Certain fumes are highly toxic and can quickly cause unconsciousness and even death if inhaled to any extent. Petrol (gasoline) vapour comes into this category, as do the vapours from certain solvents such as trichloroethylene. Any draining or pouring of such volatile fluids should be done in a well ventilated area.

When using cleaning fluids and solvents, read the instructions carefully. Never use materials from unmarked containers – they may give off poisonous vapours.

Never run the engine of a motor vehicle in an enclosed space such as a garage. Exhaust fumes contain carbon monoxide which is extremely poisonous; if you need to run the engine, always do so in the open air or at least have the rear of the vehicle outside the workplace.

The battery

Never cause a spark, or allow a naked light, near the vehicle's battery. It will normally be giving off a certain amount of hydrogen gas, which is highly explosive.

Always disconnect the battery earth (ground) terminal before working on the fuel or electrical systems.

If possible, loosen the filler plugs or cover when charging the battery from an external source. Do not charge at an excessive rate or the battery may burst.

Take care when topping up and when carrying the battery. The acid electrolyte, even when diluted, is very corrosive and should not be allowed to contact the eyes or skin.

If you ever need to prepare electrolyte yourself, always add the acid slowly to the water, and never the other way round. Protect against splashes by wearing rubber gloves and goggles.

Mains electricity and electrical equipment

When using an electric power tool, inspection light etc, always ensure that the appliance is correctly connected to its plug and that, where necessary, it is properly earthed (grounded). Do not use such appliances in damp conditions and, again, beware of creating a spark or applying excessive heat in the vicinity of fuel or fuel vapour. Also ensure that the appliances meet the relevant national safety standards.

Ignition HT voltage

A severe electric shock can result from touching certain parts of the ignition system, such as the HT leads, when the engine is running or being cranked, particularly if components are damp or the insulation is defective. Where an electronic ignition system is fitted, the HT voltage is much higher and could prove fatal.

Routine maintenance and capacities data

Oil tank	4.5 Imp pints (2.56 litres)
Gearbox	1 Imp pint (0.57 litres)
Primary chaincase	4.5 fl oz (130 cc)
Front forks	5 fl oz (150 cc)
Long scrambles type	6.5 fl oz (170 cc)
Contact breaker gap	0.014 - 0.016 in. (0.35 - 0.04 mm) - coil ignition 0.012 - 0.015 in. (0.30 - 0.38 mm) - magneto ignition
Spark plug gap	0.020 in. (0.51 mm) - magneto ignition 0.020 - 0.022 in. (0.51 - 0.56 mm) - coil ignition
Tappet clearances (engine cold): Inlet	0.003 in. (0.07 mm) - models 77, 88, 99 0.006 in. (0.15 mm) - models 'SS', 650, 750
Exhaust	0.005 in. (0.12 mm) - models 77, 88, 99 0.008 in. (0.20 mm) - models 'SS', 650, 750

Chapter 1 Engine

Contents

Specifications

Engine:

	Model 7 and Model 88 (SS)	Model 77 and Model 99 (SS)	Model 650 (SS)	Model 750
Maximum speed (approx)	90 (110)	90 (115)	115	110
Type	Twin cylinder four stroke with pushrod operated overhead valves			
Capacity	497 c.c	597 c.c.	646 c.c.	745 c.c.
Bore	66 mm	68 mm	68 mm	73 mm
Stroke	72.6 mm	82 mm	89 mm	89 mm
Compression radio				
Pre - 1960 Standard (Optional)	7.6:1 (9.0:1)	7.4:1 (8.2:1)		
Post 1960 Standard (Optional)	8.5:1 (9.45:1)	8.2:1 (9.0:1) 10.0:1 "Nomad"	9.0:1	7.5:1 (Various)
Cylinder Head				
Pre 1955	Cast Iron			
Post 1955	Aluminium Alloy	Aluminium Alloy	Aluminium Alloy	Aluminium Alloy
Cylinder Barrel	Cast Iron	Cast Iron	Cast Iron	Cast Iron
Tappet Bore Size		1.1875 in/1.1865 in		

Pistons:

Number	2
Number of rings - Compression	2
Oil Scraper	1
Ring Gap	0.009 in/0.014 in
Ring Side Clearance	0.0015 in/0.0035 in
Gudgeon Pin - 0. Diameter	0.6868 in/0.6866 in
Small End Dia	0.6878 in/0.6873 in

Valves:

Inlet head dia - Standard	1.406 in[+]	1.406 in	1.500 in	1.500 in
Inlet head dia - SS	1.406 in	1.406 in		
Exhaust head dia -	1.312 in	1.312 in	1.312 in	1.312 in
Stem Diameter - Inlet and Exhaust		0.310 in/0.3135 in.		

Rocker Clearance - Cold

Inlet - Standard (SS)	0.003 in (0.006 in)		0.006 in	0.006 in
Exhaust - Standard (SS)	0.005 in (0.008 in)		0.008 in	0.008 in

Rockers:

Bore Dia	0.5003 in/0.4998 in
Spindle - O. Dia	0.4990 in/0.4985 in
Thrust washer - width	0.015 in

Valve Springs:

Number	2 off per valve
Free length - inner	1.531 in'
Free length - outer	1.700 in'

Valve Guides:

Bore Size	0.3145 in/0.3135 in

Push Rods:

Overall length - inlet	7.928 in	8.210 in	8.210 in	8.210 in
Overall length - exhaust	7.366 in	7.366 in	7.366 in*	7.366 in*

Camshaft:

Shaft Dia's - Bearing Surfaces	0.8740 in/0.8735 in
	0.8735 in/0.8730 in
Bore of bushes	0.8750 in/0.8745 in

Crankshaft:

Journal Diameter - drive side	1.1815 in to 1.1812 in
" " timing side	1.1812 in to 1.1807 in
" " big end	1.7505 in to 1.7500 in
" " " " (500 only)	1.5005 in to 1.5000 in

Main Bearings:

Drive side	NJ306E-M1 - 30 x 72 x 19 mm
Timing side	NJ306E-M1 - 30 x 72 x 19 mm

Ignition Timing

Fully advanced	32 degrees	(30 degrees Model 88)

Spark Plug Gap

0.020 in to 0.022 in

Contact Breaker Gap	
Magneto	0.012 in
Distributor	0.015 in

Engine Sprocket

Normal	21 teeth	19 teeth (Model 88)
G15CSR	22 teeth	20 teeth (Model 99)

Chain Sizes

Primary	½'' x 0.305'' x 75 links
Rear	5/8'' x ¼'' x 97 links
" - G15CS	Mercury and Atlas
	5/8'' x 3/8'' x 97 links
- C15CSR	5/8'' x 3/8'' x 98 links (5/8'' x 3/8'' also used on post-1964 650SS Models)

+ Pre-1959 models have inlet valves of same diameter as exhaust.

' After engine number 125871 8.110 in and 7.266 in.

* Pre-1959 models have 2 inch long springs.

Torque Wrench Settings

		ft lbs
Crankshaft Assembly Nuts		25
Big-end nuts	500/600	15
	650/750	25
Oil pump nuts		12
Cylinder base nuts - 3/8 in dia		25
5/16 " "		20
Final Drive Sprocket nut		80
Clutch centre nut		70
Engine Sprocket/Rotor Nut		75
Stator nuts		15
Cylinder head nuts - 3/8 in dia		30
5/16 in dia		20
Pressure release valve		25
Engine mounting bolts		25

General Description

The "Dominator" series of engines are virtually identical over the complete range of 497cc to 745cc - excepting that dimensions vary as shown in the specifications section. Any small differences amongst the model range are mentioned in the text at the point where it affects the stripdown procedure.

From a design viewpoint, these engines are vertical twin cylinder four-strokes in which both pistons rise and fall together producing a firing impulse during every revolution of the crankshaft. The engine breathing is controlled by the pushrod operated, overhead valve layout and a timed crankcase breather.

Lubrication is effected on the dry sump principle, in which oil is fed by gravity to a gear-type pump and distributed to the various parts of the engine. A separate scavenge pump which forms part of the oil pump assembly ensures oil which drains back into the crankcase is returned to the oil tank.

The crankcase breather is situated at the top, rear of the left hand crankcase, except on 750 cc engines where it takes the form of a disc driven from the left-hand end of the camshaft.

The crankshaft assembly consists of a wide centre flywheel with two outer bob weights, one on each side. The big end bearings are of the shell type fitted to split connecting rods. The cylinder barrel is of cast iron and the cylinder head, although originally of cast iron, has been cast in aluminium alloy since 1955 on all models; both barrels and heads being monobloc castings. Ignition is provided by either a magneto or by a coil and a contact breaker assembly.

The carburation system is is dependent on the model and the year and may include single or twin carburettors on horizontal or downdraught inlet manifolds.

Before embarking on a stripdown, the author would like to point out that owing to the longevity of life of these engines and the ease of interchanging of many parts amongst the Dominator range and also the many tuning modifications available, it is possible that minor variations of the stripdown procedure given may sometimes prove necessary.

In these cases the text should be used as a guide and the owner should note the differences in procedure; for example, the machine may be fitted with flexible inlet manifolds and Concentric carburettors in place of a single Monobloc carburettor on an alloy manifold as fitted as standard.

2 Operations with engine in frame

1 It is not necessary to remove the engine from the frame unless the crankshaft assembly, the big ends or the camshaft require attention. Most operations can be accomplished with the engine in the frame, such as:

a) Removal and replacement of the cylinder head.
b) Removal and replacement of the cylinder barrel and pistons.
c) Removal and replacement of the clutch and primary drive.
d) Removal of the timing sprockets and oil pump.
e) Removal of the contact breaker and automatic advance assembly or magneto.

2 When several operations have to be undertaken simultaneously, such as during an extensive rebuild or overhaul, it is often advantageous to remove the engine from the frame.

The engine and gearbox unit are designed to be removed together and little is gained by removing the engine alone. Removal of the combined units may be achieved with the engine complete on all models, but in the case of the larger capacity models, it is advisable to remove the cylinder head to give better clearance and at the same time to lighten the load.

Removing the engine will give the advantage of better access and more working space, especially if the engine is attached to a bench-mounted stand.

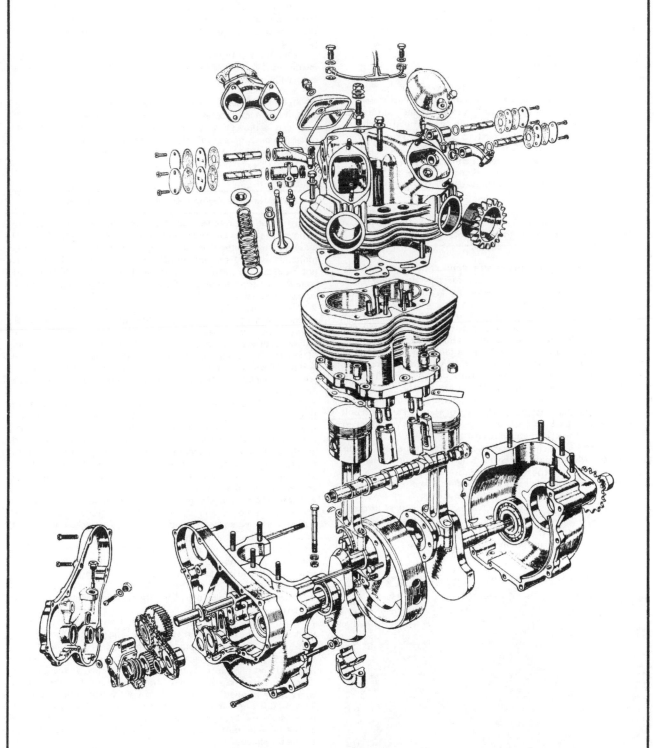

FIG. 1.1 THE EARLY 'DOMINATOR' TWIN ENGINE

3 Operations with the engine removed

1   Removal and replacement of the main bearings.
2   Removal and replacement of the crankshaft assembly.
3   Removal and replacement of the camshaft.

4 Method of engine removal

The engine is heavy, even for two pairs of hands. In consequence, it is recommended that some of the engine weight is shed before the engine is removed from the frame. Commence by removing the cylinder head along with the carburettor(s); the barrels may then also be removed to shed further weight if required. In this case, it is advisable to turn the crankshaft so that the pistons are at bottom dead centre and pack the top of the crankcase with rag to avoid damage to the pistons and connecting rods as the engine is removed.

5 Dismantling the engine - removing the petrol tank

Slimline models

1   Place the machine on the centre stand and ensure that it is standing firmly on level ground. Turn off the petrol tap(s) and disconnect the petrol pipe(s) by unscrewing the union where it joins the base of the tap. It may be necessary to hold the tap in position with a spanner whilst performing this operation. If difficulty is encountered, the petrol pipe may be disconnected at the union with the float chamber, as an alternative.
2   Remove the drain plugs from the oil tank and the crankcase, also the gearbox if this component is to be stripped. Drain whilst the engine is warm so that the oil runs out more freely. Note that the crankcase drain plug contains a filter, which should be removed and cleaned in petrol. It is retained by a large circlip.
3   Unlock the Dzus fastener at the rear of the dual seat and pull the seat up and back to clear the two locating pegs which retain the front of the seat, and locate it laterally.
4   Remove the left hand side cover to expose the battery. Undo the two nuts which retain the battery clamp, pull out the battery and disconnect it at the terminals. Store the battery in a safe place after taking the opportunity to check and re-charge it if necessary.
5   The petrol tank is secured by two bolts which pass through plates welded to the inside of the top frame tubes, forward of the cylinder head. Remove these bolts complete with the metal washer and rubber spacers (both sides of the frame).

5.1 Disconnecting the petrol pipe at the carburettor

6   The back of the tank rests on rubber blocks, wrapped round the frame, and is retained by a rubber ring. Remove the ring and pull the tank up and back to clear the steering head. Ensure that the rubber blocks are not misplaced or annoying vibration will be transferred to the tank when the engine is replaced and running.
Note the position of the rubber support pads over the main frame tubes so that they are located accurately when the tank is eventually replaced.

Since there is no necessity to drain the tank prior to these operations, it can represent a dangerous fire hazard. Make sure it is placed well away from the machine and any naked flames or other sources where involuntary ignition may occur.
Also, ensure that the taps do not leak petrol.

6 Dismantling the engine - removing the cylinder head

1   In order to preserve a leaktight joint in the exhaust system, remove the exhaust pipe and silencer as a complete unit. Unscrew the finned exhaust pipe retaining rings, making judicious use of a flat-nose punch and hammer if a 'C' spanner is not available. If difficulty is encountered, it is recommended that the Norton service tool 06-3968 "C" spanner - is obtained, or a replica made.
2   Many types of silencer may be obtained for the Dominator engines. In general, if the engine has the standard low level or swept back type pipes, the silencer will be mounted by the pillion footrest to an extension of the main frame loop. "Dunstall" type silencers require an additional strut from the end of the silencer to the upper frame extension. High level pipes are also available, the mounting of the silencers being dependent on the type of installation.
3   Some machines are fitted with 'Siamese' exhaust pipes. In this case, remove the two finned rings and the offside pillion footrest. Finally, some owners may have fitted exhaust pipes which have an interconnection balance pipe. A clip will be found on the balance pipe - undo this and split the exhaust system. In all cases, try to avoid dropping the exhaust system on the floor as it is removed because dents and scratches will mar the appearance of these items.
4   Note there is a sealing ring in each exhaust port, which should be removed and discarded. It is customary to fit new replacements when the exhaust system is eventually refitted in order to preserve a leaktight joint.
5   Remove the engine steady plate by taking out the two nuts and bolts which hold the plate to the frame and unscrew the nut from the stud in the cylinder head. Also, disconnect the horn if fitted to this plate.
6   Disconnect the plug caps from the sparking plugs and remove the latter.
7   Unscrew the two large bolts which transmit the rocker oil feed. Special care is needed when removing the oil feed pipe to prevent the thin copper pipe from necking or twisting. If this should happen, it is recommended that the whole pipe is removed and sawn in two, using a jewellers saw, at the point of deformation. The pipe may then be joined (after being thoroughly cleaned) by means of a tightly fitting flexible pipe. This modification has been extensively used where vibration is responsible for breaking pipe/union joints. Disconnect the pipe at the flexible joint, if possible, or leave the feed pipe attached to the timing cover.
8   Disconnect the air cleaner (if fitted) from the carburettor by removing the flexible hose connecting the carburettor to the air filter stored between the oil tank and the battery cover.
9   Remove the throttle and slide assembly of each carburettor fitted by unscrewing the large ring nut from the top of Monobloc carburetters; or in the case of Concentric carburetters, by unscrewing the two 2BA slotted screws retaining the top. Pull out the throttle slide assemblies and tape them to the frame, out of harm's way, after wrapping them in clean rag

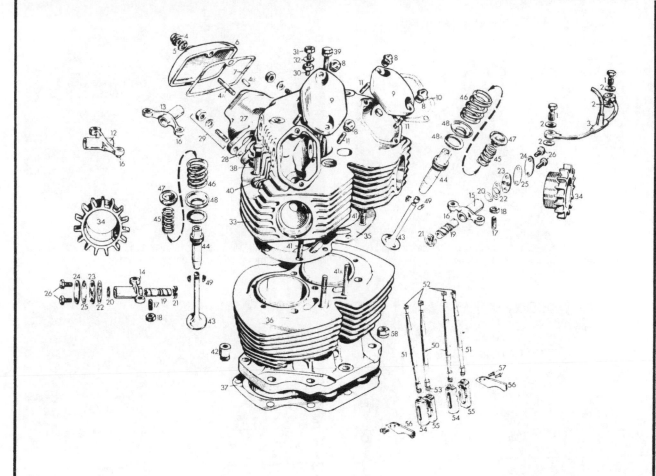

FIG. 1.2 CYLINDER BARREL AND HEAD ASSEMBLIES

1 Banjo bolt - oil feed (2 off)
2 Fibre washer (4 off)
3 Rocker oil feed pipe assy
4 Domed nut - rear rocker box
4A Central stud - rear rocker box
4B Locating pin - rear rocker box
5 F/Cu washer
6 Rocker box cap - rear
7 Rocker box gasket - rear
8 Domed nut - front rocker box (4 off)
9 Rocker box cap - front (2 off)
10 Rocker box gasket - front (2 off)
11 Rocker box mounting stud - front (4 off)
12 Rocker inlet - right hand
13 Rocker inlet - left hand
14 Rocker exhaust - right hand

15 Rocker exhaust - left hand
16 Rocker ball end (4 off)
17 Rocker adjusting screw (4 off)
18 Adjuster locking nut (4 off)
19 Rocker shaft (4 off)
20 Shaft thrust washer (4 off)
21 Shaft spring washer (4 off)
22 Inner gasket (4 off)
23 Locking plate (2 off)
24 Outer plate (4 off)
25 Outer gasket (4 off)
26 Rocker plate retaining bolt (8 off)
27 Inlet manifold (single carb, type)
28 Manifold/head gasket (2 off)
29 Stud, nut, washer manifold mounting
30 Stud in head - steady plate

31 Stud nut
32 Stud washer
33 Cylinder head
34 Exhaust pipe locking ring (2 off)
35 Cylinder head gasket
36 Cylinder barrel
37 Cylinder base gasket
38 Cylinder head bolt - long (4 off)
39 Cylinder head bolt - short (1 off)
40 Washer - (4 off)
41 Stud - Cylinder head to barrel (3 off)
41A Stud - barrel to head (2 off)
42 Sleeve nut - front stud (2 off)
43 Exhaust valve (2 off)
44 Valve guide (4 off)
45 Valve spring - inner (4 off)
46 Valve spring - outer (4 off)

47 Valve spring top cup (4 off)
48 Valve Spring bottom collar (4 off)
48A Heat insulating washer (4 off)
49 Valve collett (4 pairs)
50 Push rod assembly - inlet (2 off)
51 Push rod assembly - exhaust (2 off)
52 Push rod top (4 off)
53 Push rod bottom (4 off)
54 Right hand tappet (2 off)
55 Left hand tappet (2 off)
56 Tappet location plate (2 off)
57 Plate screw (4 off)
58 Cylinder base nut and washer (9 off each)

6.1 "Injudicious" use of a drift and hammer

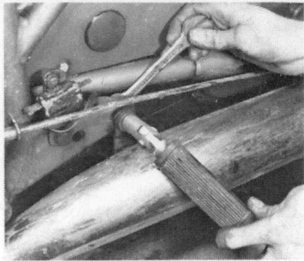
6.2 Silencer mounted by the pillion footrest

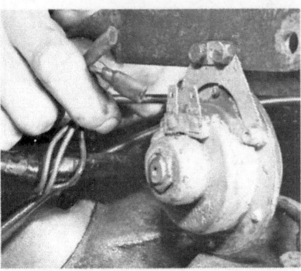
6.5a Pull "Lucar" connectors from the horn

6.5b Engine steady between steering head and cylinder head

6.7 Unscrew rocker feed banjo bolts
Note: Alloy or copper washers each side of union

6.9a Throttle and slide assembly - Monobloc carburettor

or clean plastic bags. The carburettor(s) and manifold(s) should also be removed.

10 Inlet manifolds may be of the following types:

a Single carburettor - vertical studs, 2 off per intake
b Twin carburettors - solid manifold with vertical studs - 2 off per intake
c Single carburettor - horizontal studs - 4 off per intake (the two inner studs will, in fact be Allen screws. which are not accessible until the carburettor body is removed)
d Twin carburettors - horizontal studs - 4 off per intake
e Twin carburettors - individual induction spacers with horizontal studs - 4 off per intake
f Twin carburettors - flexible mounting.

11 On coil ignition models, it may be necessary to remove the ignition coil from its position below the petrol tank, run back in order to remove the cylinder head or to gain access to the cylinder head bolts.

12 Remove the three rocker covers. The cylinder head is retained by a total of five bolts and five nuts. Slacken and remove nine of them, leaving only the front centre bolt in position. This, like the other two nuts on either side, will require a slim socket or box spanner for its release. With regard to the other nuts and bolts, two nuts will be found on the underside of the cylinder head, recessed into the fins at the front of the engine. Another nut is found on the underside of the cylinder head, at the rear. The four bolts remaining are easily accessible from the top of the cylinder head. Depending on the model, the spanners required for removing the nuts located on the underside of the cylinder head should be selected from the following:

a A flat ring or open-ended spanner for the rearmost nut
b A slim socket or box spanner as used for the front centre bolt for the two nuts recessed into the fins at the front of the engine.

13 When the front cylinder head bolt is removed last of all, the cylinder head will tilt a little against the spring pressure of the valve that is open. This will obviate the need to break the cylinder head joint. If for any reason the joint does not separate of its own accord, ease it away with a wooden drift under the exhaust and inlet ports. As a last resort, replace the plugs and gently turn the engine over using the kickstart lever.

6.9b Two nuts retain carburettor to the manifold
Note: heat resistant spacer

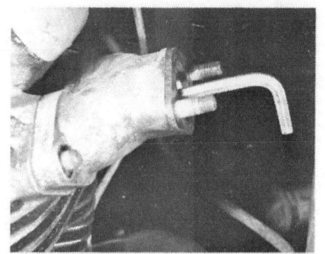
6.9c Allen screws inside the manifold must be removed on single carb - horizontal stud type manifolds

6.10a Single domed nut retains inlet rockers cover

6.10b Two domed nuts retain each exhaust rocker cover

All Engines - Featherbed Frame

14 Before the cylinder head can be lifted away, it is necessary to feed each of the four pushrods into the cylinder head as far as possible, after detaching them from the ends of the rocker arms. This can be accomplished by tilting the cylinder head towards the rear, whilst holding the cylinder head with one hand and the pushrods with the other. Do not use force and make sure the pushrods are clear of the cylinder barrel as the head is being removed. Failure to observe this precaution may cause damage to the light alloy pushrods, necessitating their renewal. Sometimes it is convenient to lift the cylinder head and barrel together, after all the nuts and bolts have been removed.

Atlas Engine - Scrambles Frames

15 In addition to the above note, it is also necessary to remove the exhaust rockers in order that the pushrods may be fed further into the head - see paragraph 16.

Note: On featherbed frames, the head should be turned through 90° on removal and extracted through the top frame tubes.

Other models (ie. Models 7 and 77) No problems should be encountered in lifting the head clear of the pushrods.

16 The cylinder head gasket will adhere to either the cylinder barrel or cylinder head and should not be re-used unless it is completely undamaged.

17 On removing the head, take out the push rods and mark them so that each push rod is replaced in the position it previously occupied.

18 In order to extract the rocker spindles, a bolt about 1'' long, 5/16 x 26 tpi is required along with a washer and a piece of tube of about ¾'' in length and inside diameter of about 0.5''. Place the washer against the head of the bolt and slide the tube over the threads against the washer.

19 Remove the two bolts, two gaskets and two plates which retain each rocker spindle (note their order of assembly). This will expose the rocker spindle which has a threaded inside diameter. Screw the 5/16'' bolt into this hole and align the spacer tube centrally round the spindle bush. Screw in the bolt and the spindle will be extracted into the tube. As the bolt is screwed in, ensure that the spring washer and the thrust washer on each rocker arm do not fall down the push rod tunnels. Extract the rocker arm after removing each spindle. Note: If the rocker spindles require force to extract, soak the head in hot water to facilitate removal. Some are a very tight fit.

6.14a Lift cylinder head vertically to clear studs

6.14b Tilt head to the rear and feed pushrods into head

---

7 Dismantling the engine - removing the cylinder barrel and pistons

---

1 The cylinder barrel is retained to the crankcase by nine studs. Remove the nine holding down nuts by unscrewing the nuts until they touch the bottom fin and then lift the barrel slightly and remove all the nuts. If broken piston rings are suspected, then put the engine at T.D.C. and pack the top of the crankcase with rag as the barrels are withdrawn: This will also help to support the pistons as they emerge from the bore.

2 Remove and discard the circlips from each piston, then remove both gudgeon pins, taking care to support the piston and connecting rod as they are tapped out of position. If the pins are a tight fit, the pistons should be warmed first by placing a rag soaked in hot water on each crown. This will expand the alloy of the piston and release the grip of the gudgeon pin boss. On no account use force, or the connecting rods may be damaged permanently.

3 Mark each piston INSIDE the skirt to ensure it is replaced in its original position. The pistons are individually marked on the crown to this effect, but the marks may have been erased if the crown was badly scratched during a previous decoke. The 750 cc Atlas models have valve cutaways, the cutaway for the exhaust valve positioned much nearer to the outer edge. This makes the correct location of the pistons vital; ie. the distance between the exhaust valve cutaways is greater than the distance between the inlet valve cutaways.

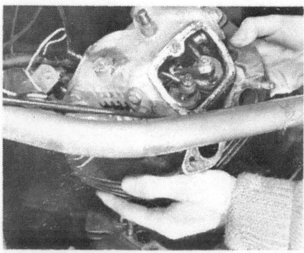

6.14c Turn head thr' 90° and extract from frame

FIG. 1.3 CRANKSHAFT ASSEMBLY AND TIMING GEAR

1 Crankshaft
2 Drive side roller main bearing
3 Woodruff keys
4 Timing side ball main bearing
5 Oil thrower
6 Triangular washer
7 Half-time pinion
8 Oil pump
9 Sealing washer
10 Connecting rod - 2 off
11 Bolt - 4 off
12 Big-end shell bearing - 4 off
13 Big-end cap - 2 off
14 Nut - 4 off
15 Piston - 2 off
16 Gudgeon pin - 2 off
17 Circlip - 4 off
18 Camshaft
19 Woodruff key
20 Camshaft sprocket
21 Nut
22 Camshaft chain
23 Intermediate pinion and sprocket
24 Magneto drive chain
25 Bush
26 Thrust washer
27 Spindle
28 Inner tensioner plate
29 Camshaft chain tensioner
30 Outer tensioner plate
31 Nut - 2 off
32 Spring
33 Timing breather rotary plate
34 Timing breather stationary plate

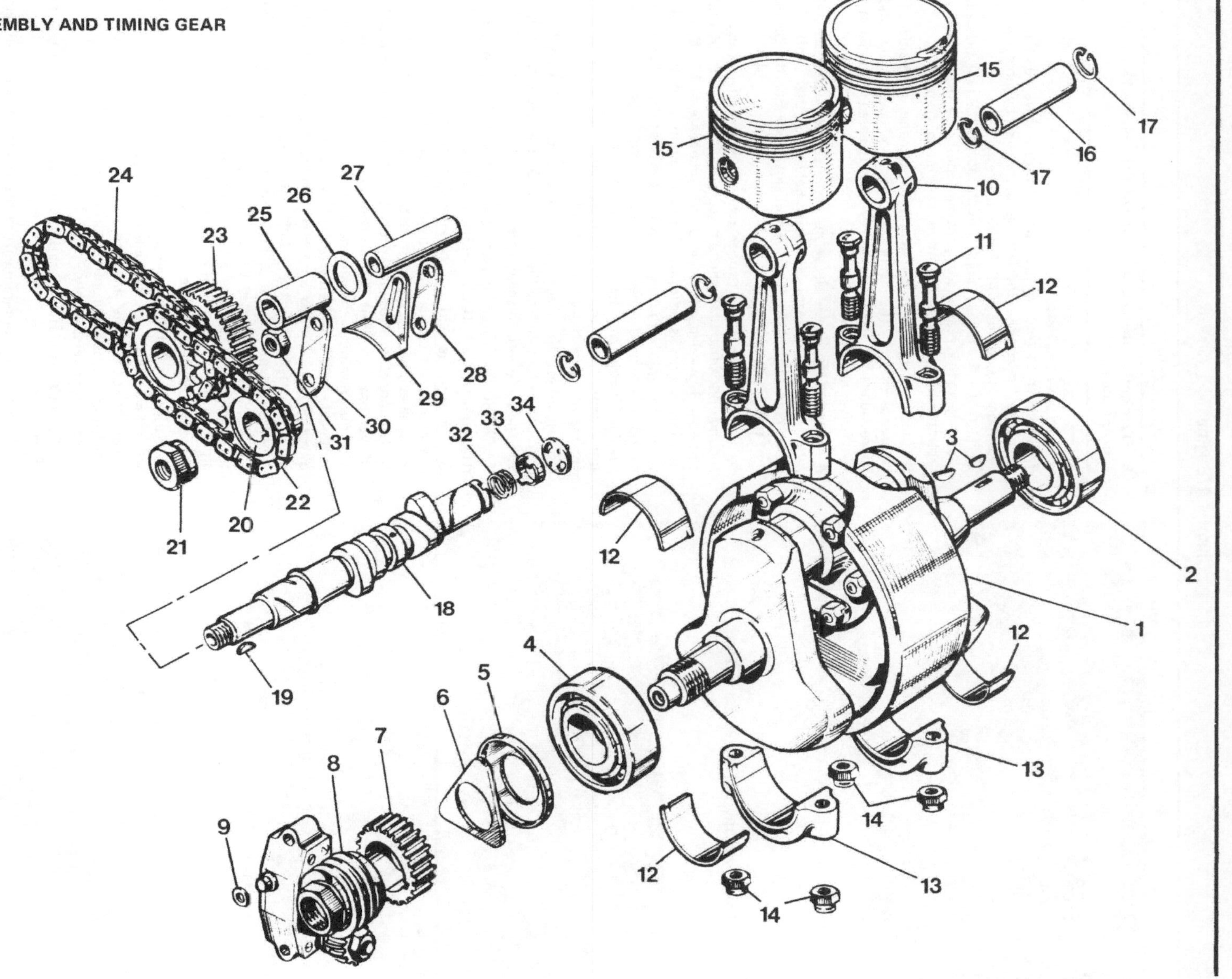

7.1a Lift the barrels slightly to remove the retaining nuts

7.1b The barrels have long spigots which locate with the crankcase

7.2 Pad crankcase mouth before removing circlip

8 Dismantling the engine - removing the alternator, clutch and primary chaincase

Pressed steel chaincase

1 In order to remove the outer primary drive case, it is first necessary to remove the left hand footrest by unscrewing the retaining domed nut. Pull off the footrest.

2 Remove the grease nipple which locates the rear brake lever; disconnect the rear brake arm and pull the rear brake lever assembly off its support. Remove the return spring.

3 Place a large tray immediately below the underside of the chaincase joint. The chaincase has no drain plug and in consequence the full oil content will be released immediately the two halves are separated.

4 Remove the centre sleeve nut, washer and rubber seal which retain the two primary chaincase halves together. Pull off the outer cover from the forward end slowly so that the oil drains into the tray provided.

5 Pull off the outer rubber sealing band and inspect it for rotting and splitting. If the chaincase leaks it may be the fault of this seal. However, the pressed steel chaincase as fitted to most of these engines is notorious for leaking and the only permanent cure is to replace the casings with new ones should any damage or buckling cause excessive leakage of oil: the alternative is a regular check of the oil level and frequent removal and immersion of the primary chain in a chain lubricant to preserve the life of the primary drive assembly.

6 Disconnect the leads from the alternator (if fitted) at the snap connectors, noting the colour coding across each connector. Pull the cable into the chaincase.

7 Remove the three nuts and star washers holding the stator to its mounting. Pull off the stator and leads.

8 The rotor is held on a parallel shaft by a nut on the end of the crankshaft and is located with a Woodruff key. In order to lock the engine, place a stout metal rod through the small end of both connecting rods and rotate the engine in an anticlockwise (drive) position until the rod rests across the crankcase mouth. The nut retaining the alternator rotor may be slackened and removed. Use a socket wrench and turn in an anti-clockwise direction; the nut has a normal right hand thread. Do not misplace the washer which seats below the nut.

   Alternatively, if the cylinder barrel has not been removed, the primary drive may be locked by having a second person to hold on the rear brake with a ring spanner, with the engine in 4th gear. Pull off the rotor with the help of a sprocket extractor, if required - retain the Woodruff key

9 Strip the clutch by removing the three slotted nuts using a screwdriver. Pull out the clutch springs and spring cups, the outer pressure plate and the plain and friction plates. When removing the plates from the drum, the original order should be preserved for re-assembly.

10 Pull out the clutch operating pushrod from the gearbox mainshaft.

11 Remove the spring link from the primary chain and remove the chain. (This operation may require the chain to be slackened by moving the gearbox and possibly the back wheel forward). Pull off the clutch outer drum.

12 With the back wheel still locked in gear, remove the nut and spring washer from the gearbox mainshaft which retains the clutch centre. Pull the centre off the splined shaft - if difficulty is encountered EXTREME care must be taken when using an extractor to ease the centre off its shaft otherwise damage will occur to the centre and/or the casing. A special service tool extractor is available.

13 The engine sprocket is held on a taper and also located with a Woodruff key. Ease the sprocket off with a sprocket extractor - if difficulty is encountered, the sprocket may be lightly warmed around its central boss to break the taper.

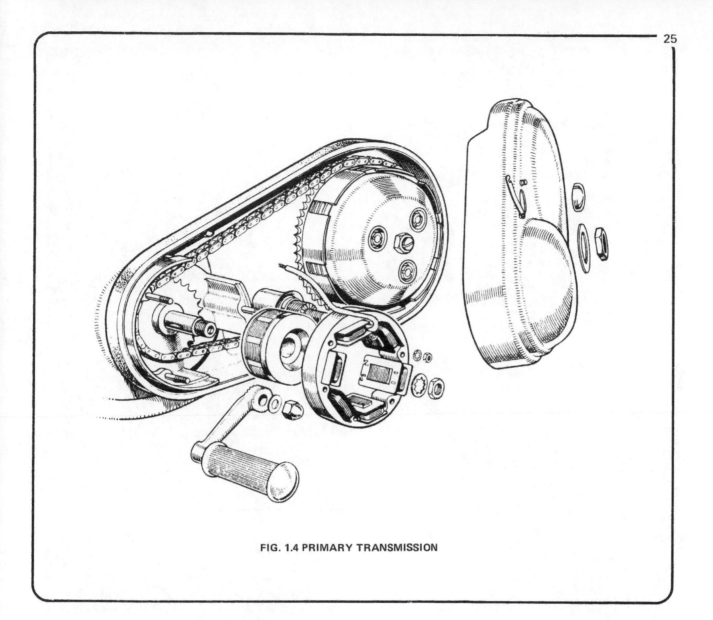

FIG. 1.4 PRIMARY TRANSMISSION

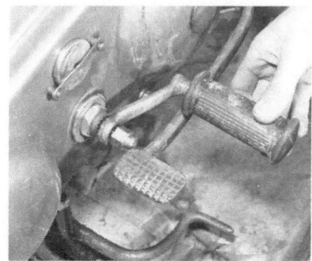

8.1 Left hand footrest retained by chromed nut

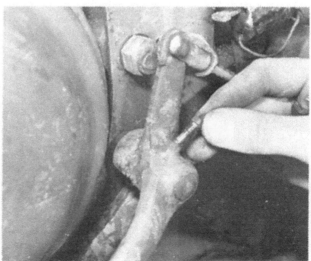

8.2 Grease nipple retains brake pedal

8.8 Nut and flat washer retain rotor

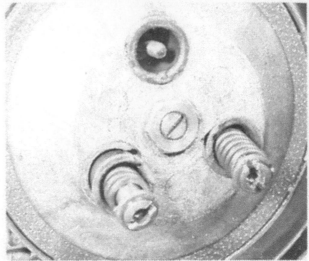

8.9a Remove the slotted clutch spring adjusting nut, spring and cup

8.9b Pull out the plain and friction plates

8.10 Removing the clutch push rod

8.11 Clutch outer drum pulls off after pressure plate and clutch plates are removed

8.12 Clutch centre assembly is splined to the gearbox mainshaft

8.15 The inner primary chaincase

8.17 Nut retained by star washer and small screw

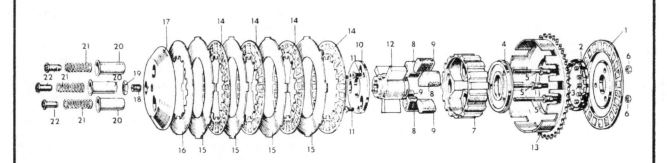

FIG. 1.5 CLUTCH ASSEMBLY*

1 Clutch backplate (bonded)
2 Clutch roller cage
3 Clutch rollers - 15 off
4 Clutch race plate
5 Clutch spring stud
   3 off
6 Clutch spring stud nut
   3 off
7 Clutch body

8 Clutch shock absorber
   rubbers (large) 3 off
9 Clutch shock absorber
   rubbers (small) 3 off
10 Clutch shock absorber
    cover plate
11 Screws for item 10
    3 off

12 Clutch shock absorber
    centre
13 Clutch sprocket
14 Clutch friction plate
    (bonded) 5 off
15 Clutch plate (plain)
    5 off
16 Clutch end plate
    assembly (bonded)

17 Clutch pressure plate
18 Clutch adjuster
19 Locknut for item 18
20 Clutch spring cup
    3 off
21 Clutch spring 3 off
22 Clutch spring
    adjusting nut 3 off

*early models have the clutch plates assembled in reverse order

14 Three countersunk Allen screws hold the alternator housing to the crankcase. Remove these screws and pull off the housing. Also, remove the three cheese head screws which hold the inner chaincase to the crankcase.

15 The inner chaincase may now be removed by first withdrawing the bolt which holds the upper half of the casing to a spacer from the engine plates and then releasing the bracket which is attached to the large stud which locates the bottom of the gear box.

16 Remove the sleeve spacer which engages with the engine plates against which the inner casing and the footrest mate.

Cast alloy chaincase - scramble frame models only

17 Removal of the aluminium covers as employed on the G15 and N15 models closely follows the above procedure, which should be used as a guide. The only major exception is that the outer cover is retained around its periphery by 14 screws in place of the centre nut.

With the back wheel still locked in top gear, loosen the cheese headed screw which retains the plate locating nut on the gearbox mainshaft. Remove the washer and them remove the nut which has a LEFT HAND thread.

18 Turn the backwheel until the split link in the rear chain is situated on the rear sprocket. Remove the split link and the rear chain. Pull off the final drive sprocket with the aid of a puller if necessary.

19 Disconnect the earth lead on magneto models or the low tension lead on coil ignition models.

9 Dismantling the engine - removing the crankcase assembly from the frame

1 As mentioned previously, the engine and gearbox are designed to be removed as a unit. This is recommended on two accounts. Firstly, it is quicker and easier to accomplish and, secondly, the gearbox/engine plates/crankcase assembly provides a convenient supporting base whilst the inside of the timing cover is stripped.

2 Unscrew the tachometer drive cable (if fitted) from the gearbox mounted on the timing cover. Also, remove the rocker feed pipe (late models) which is joined to the timing cover by a banjo union. Take care not to lose the sealing washers.

3 Remove the gear indicator from the gearbox end cover which is retained by a setscrew and the gear change lever, secured on splines by a pinch bolt. Push the crankcase breather pipe clear of its mounting on the crankcase. Note: This breather may be vented to either the oil tank or to lubricate the rear chain or, on older machines, may be vented to atmosphere. In the case of the breather lubricating the rear chain, observations should be made periodically to check that excess oil is not "lubricating the rear tyre". If this should happen, engine damage is likely, which requires urgent attention.

4 Remove the right-hand footrest complete with spacer and through bolt; it should at this stage pull out but may require a few turns of the securing nut to release, especially if bent.

5 Remove the single bolt which attaches the oil pipes to the crankcase. Also, remove the oil tank; this is retained by two nuts underneath and a nut, bolt and rubber washer at the top. The rubber washer and the rubber mounting pad on which the tank sits isolate the tank from vibration. Disconnect the chain breather from the oil tank to the chainguard.

6 Remove the battery case, pull out all electrical leads and release the two screws in the bottom and the nut and bolt at the top of the casing. On coil ignition models, where the rectifier is mounted on the battery case, disconnect the terminals, noting the colour coding.

7 Screw in the clutch cable adjuster at the gearbox, remove the clutch cable inspection plate (2 screws) and disconnect the clutch cable from the forked arm within the gearbox outer cover.

8 The stand spring is mounted on a stud fixed to the inside of the left hand engine plate. Lever the spring off this stud after removing the plate across the top of the engine plates (if fitted). Having done this, care should be taken not to knock the bike off its stand, and the owner may prefer to place a box beneath the bottom frame loop to support the machine and remove the stand (for inspection) by unscrewing the nut and bolt that passes through the centre of each of the hollow centre stand nuts and bolts to lock them, then the nut and bolt which retains each side of the stand to the frame.

9 Before removing any engine bolts, loosen all those which require extracting ie. the two pairs of nuts and bolts which retain the front engine plates to the frame; the bottom front frame/engine plate/crankcase pickup; the RIGHT HAND NUT of the top front frame/engine plate pickup; the four pairs of nuts and bolts which locate the gearbox plates to the cross members of the frame at the gusset plates. Remove the plate above the front engine plates (if fitted).

10 Utilising two people - one either side - lever the engine unit up and down, using a length of wood, in order to help remove the necessary bolts as per the order in previous paragraph noting the positions of the plain and spring washers.

11 Allow the engine to settle on the bottom frame loop whilst removing the rear bolts. Grasping the kickstart and timing case on the right hand side and the gearbox mainshaft and crankshaft on the left hand side, lift the engine unit up and out to the right-resting it on the frame loop as the left hand helper comes round the bike.

G15CS

The removal of this unit is identical to the above except that the engine unit must be raised forward, up and out to the right Also, if the head and barrel have not been removed, the engine will require tilting to the right as it is lifted.

10 Dismantle the engine - removing the magneto/distributor and auto advance assembly

1 In order to remove the ignition unit, it is first necessary to remove the twelve screws (ten on earlier models) which retain the timing cover in position. As the cover is removed, oil will leak out. If necessary, ease off the cover with a rawhide mallet, tapping behind the pressure release valve and on the opposite side of the timing case. Two dowels locate this cover - remove these if loose. Also, remove the oil seal on the oil pump if loose. Finally, note the positions of the long and short screws which retain the cover - store the screws in position by pushing them through a shoe box lid on which an outline of the timing case has been drawn.

9.2a Rocker feed banjo attachment to timing cover. Note footrest mounting in background

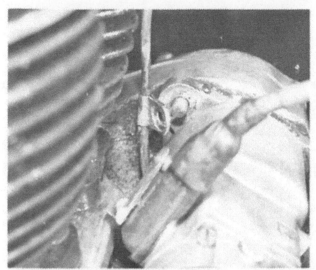

9.2b Clip retaining rocker feed to magneto stud

9.3 Crankcase breather (750 c.c. models)

9.5a The oil tank has two studs projecting down

9.5b Rubber mounting on upper oil tank bracket

9.7 Access is gained to clutch cable by removing cover

9.8a The stand return spring between the engine plates

9.8b Nut and bolt retain stand

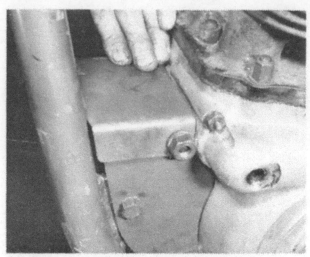

9.9a Loosen top engine bolt to remove the upper cover

9.9b Lower rear mounting bolts. Note: primary chaincase attaches to large gearbox stud.

9.9c Showing front engine plates removed and 4 nuts and bolts removed at rear - unit ready for removal

2 Magneto

The magneto is held to the crankcase by 2 studs at the top and a nut and bolt through the bottom. Before loosening these nuts, the advance/retard mechanism must be removed; this unit is retained by a central, self-extracting bolt. Simply undo the nut and the sprocket assembly will be extracted; it may be necessary to lock the engine during this operation. The nut will slacken initially, then tighten again as it commences to draw the unit off the tapered shaft.

3   Remove the three nuts and pull out the magneto complete with HT leads.

4 Coil Ignition

The distributor driving sprocket is located by a parallel pin passing through the distributor shaft and pinion; the parallel pin is retained by a circlip encircling the boss on the pinion. Remove the circlip and tap out the pin. Pull the sprocket off the end of the distributor.

6   Behind the sprocket is a tubular spacer and copper washer

(early models) or a bronze thrust washer. (later models). It should be removed and retained.

6   The distributor body is removed by slackening the clamping bolt and pulling it out of position.

7   Pull out the distributor assembly complete ie. HT leads, cover, contact breaker and advance/retard unit.

8   The distributor mounting flange may be removed from the crankcase by removing the single set screw holding the clamping flange to the inside of the timing cover extension (early models); or by removing the 2 nuts and the bolt which attach the triangular flange to the timing cover extension (later models).

9   On very early models, the dynamo should be removed at this stage. Remove the retaining nut and washer, so that the fibre drive pinion can be drawn off the dynamo drive shaft. Disconnect the two-pin electrical plug in the end cover by unscrewing the retaining screw and release the strap around the dynamo body. A stud projects from the other end of the dynamo body, through the timing cover, to locate the dynamo drive correctly. Remove the nut from the end of this stud and pull the dynamo from the back of the timing case.

11 Dismantling the engine - removing the oil pump and timing chain

1   The oil pump may be a tight fit on the two mounting studs and is best released by rotating the oil pump drive pinion so that it will travel along the worm drive of the crankshaft and ease itself off the studs.

2   The drive worm is integral with the left hand thread nut on the end of the crankshaft. Lock the engine by placing a stout metal bar through the small ends of both connection rods, then slacken the nut in a CLOCKWISE direction with a socket spanner. The nut is tight and some force may be necessary to start it moving.

3   Whilst the engine is still locked in position, slacken the nut retaining the camshaft sprocket. This has a normal right hand thread and it will be necessary to re-arrange the method of locking the engine since this nut will turn in the opposite (anti-clockwise) direction. Use steady and not excessive force and a cut-away timing cover to prevent damage to the intermediate gear spindle. Note: On models fitted with a tachometer, this nut is replaced by a special nut with a slotted drive.

4   Loosen the two nuts retaining the timing chain tensioning slipper until the slipper is free to move.

5   Because the timing chain is of the endless variety, the two sprockets and chain must be removed in unison. If the camshaft sprocket is a little tight on the end of the camshaft, it can be eased off with a pair of tyre levers, taking care not to bruise the sealing face of the timing cover joint by placing rag on the engine faces. Note: The camshaft sprocket is retained by a Woodruff key.

6   Remove the chain slipper complete with the two plates and pull off the thrust washer, also the shaft if loose.

The timing of the engine will be eased on assembly if the relative positions of the sprockets is retained. This may be achieved by pinching each chain run together between both pairs of sprockets and wiring them together.

12 Dismantling the engine - removing the crankshaft pinion and separating the crankcases

1   The crankcase pinion - or half-time pinion may be withdrawn with the use of a sprocket puller with very thin jaws. If difficulty is encountered in obtaining a grip for the puller on the sprocket, Norton service tool 067524 should be used.

2   When the pinion has been withdrawn the Woodruff key and backing washer can be lifted off the crankshaft. There is a lipped oil sealing disc which tends to cling to the right hand main bearing due to the residual oil film. This washer is best removed by two small magnets.

3   The crankcase assembly must now be removed from the engine plates. Three bolts hold the crankcase to the main engine plates - ensure that the RIGHT HAND nut is removed in the case of the top bolt. When the bolts are released, the footrest spacer should drop out.

4   The crankcase halves are retained by two screws in the bottom protrusion, two nuts on studs fore and aft of the crankcase mouth, a short bolt and nut at the front and a long bolt and nut at the bottom of the crankcase assembly.

When all of these fixings have been removed, part the crankcases by screwing a long ¼ inch Whitworth bolt into one of the bottom holes and hitting it whilst holding the right-hand crankcase. It will be noted that as the left-hand crankcase is lifted away, the rotary breather disc and spring will be displaced from the camshaft bush (left hand side).

5   Withdraw the camshaft from the right hand crankcase.

6   It is preferable to ease the right hand crankcase from the crankshaft rather than vice versa. (See Fig. 1.7) The most effective means, which obviates striking the end of the crank-shaft, takes the form of a spacer in the form of a hollow tube which abuts against the crankshaft shoulder. If the crankcase is

10.1 Remove oil pump/cover seal

10.2 Advance-retard mechanism has self extracting bolt

10.3 Two studs and a nut and bolt at the bottom hold the magneto

11.1 Pull the oil pump off its two locating studs when clear of the worm drive

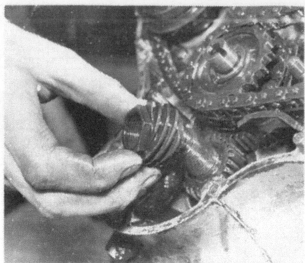

11.2 Integral worm drive and nut have LEFT HAND thread

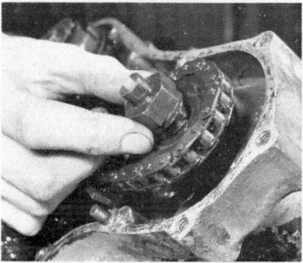

11.3 Camshaft sprocket retaining nut slotted for tachometer drive

11.4 Timing chain tensioner with both retaining nuts removed

11.5 Removing the timing drive complete

12.1 Crankshaft pinion is located by Woodruff Key

12.3 Three bolts attach crankcases to engine plates

12.4 Crankcases are retained by two screws in the bottom protrusion ...

12.4b ... and two nuts and bolts, one long at the bottom and one short in front ...

rested on the workbench, inner face uppermost, the spacer will permit the crankcase to be driven off the crankshaft by hammering downwards on a block of soft wood pressed against the crankcase casting. This is an operation requiring the assistance of a second person, one to hold the crankcase and crankshaft assembly and one to drive the crankcase in a downwards direction. The alternative is to heat the crankcase so that the right hand main bearing is displaced with the crankshaft assembly, to be drawn off at a later stage, if desired.

7 The camshaft bushes need not be removed unless wear necessitates their renewal. The bushes have an extremely long life under normal service and require special machining facilities when renewal is necessary, a task requiring expert attention.

8 The main bearings are best removed from the crankcases by heating the area around the bearing housing and bumping the jointing face against a flat wooden surface so that they are displaced by the shock. Use only a domestic oven for applying heat.

13 Examination and renovation - general

1 Before examining the parts of the dismantled engine for wear, it is essential that they should be cleaned thoroughly. Use a petrol/paraffin mix to remove all traces of old oil and sludge that may have accumulated within the engine.

2 Examine the crankcase castings for cracks or other signs of damage. If a crack is discovered, it will require specialist repair, or the renewal of both crankcases. Crankcases are supplied in matched pairs since it is considered bad engineering practice to renew only one. Under these latter circumstances there can be no guarantee that the main bearing housings have been bored exactly in line with one another.

3 Examine carefully each part to determine the extent of wear, if necessary checking with the tolerance figures listed in the Specifications section of this Chapter. The following sections of this Chapter describe how to examine the various engine components for wear and how to decide whether renewal is necessary.

4 Always use a clean, lint-free rag for cleaning and drying the various components prior to reassembly, otherwise there is risk of small particles obstructing the internal oilways.

14 Crankshaft, big end and engine bearings - examination and renovation

1 Check the big end bearings for wear by pulling and pushing on each connecting rod in turn, whilst holding the rod under test in the vertical plane. Although a small amount of side play is permissible, there should be no play whatsoever in the vertical direction if the bearing concerned is fit for further service.

2 The big end bearings take the form of shells. To gain access to the shells, remove the connecting rods by unscrewing the two self locking nuts at each end cap. If the alignment marks scribed across one face of the rod and cap cannot be seen, make new marks to ensure correct reassembly. When the nuts have been withdrawn completely, the connecting rod and end cap can be pulled off the crankshaft, with the bearing shells still attached. Mark both the connecting rods and their end caps clearly so that there is no possibility of them being interchanged. Note that the locating tabs of the bearing shells fit to the same side of each connecting rod.

3 If the crankshaft assembly is to be separated, it it advisable to continue the dismantling in a metal tray. The assembly holds approximately one teacup of oil which will be released when the cranks are separated from the centre flywheel.

4 Slacken the nuts on the right hand (timing) side of the crankshaft assembly, noting that two of the nuts are retained by a tab washer and the other four have been centre-punched to prevent them from loosening in service. All will be tight.

5 Mark the central flywheel so that there is no possibility of it being reversed during subsequent reassembly. Jar the crank

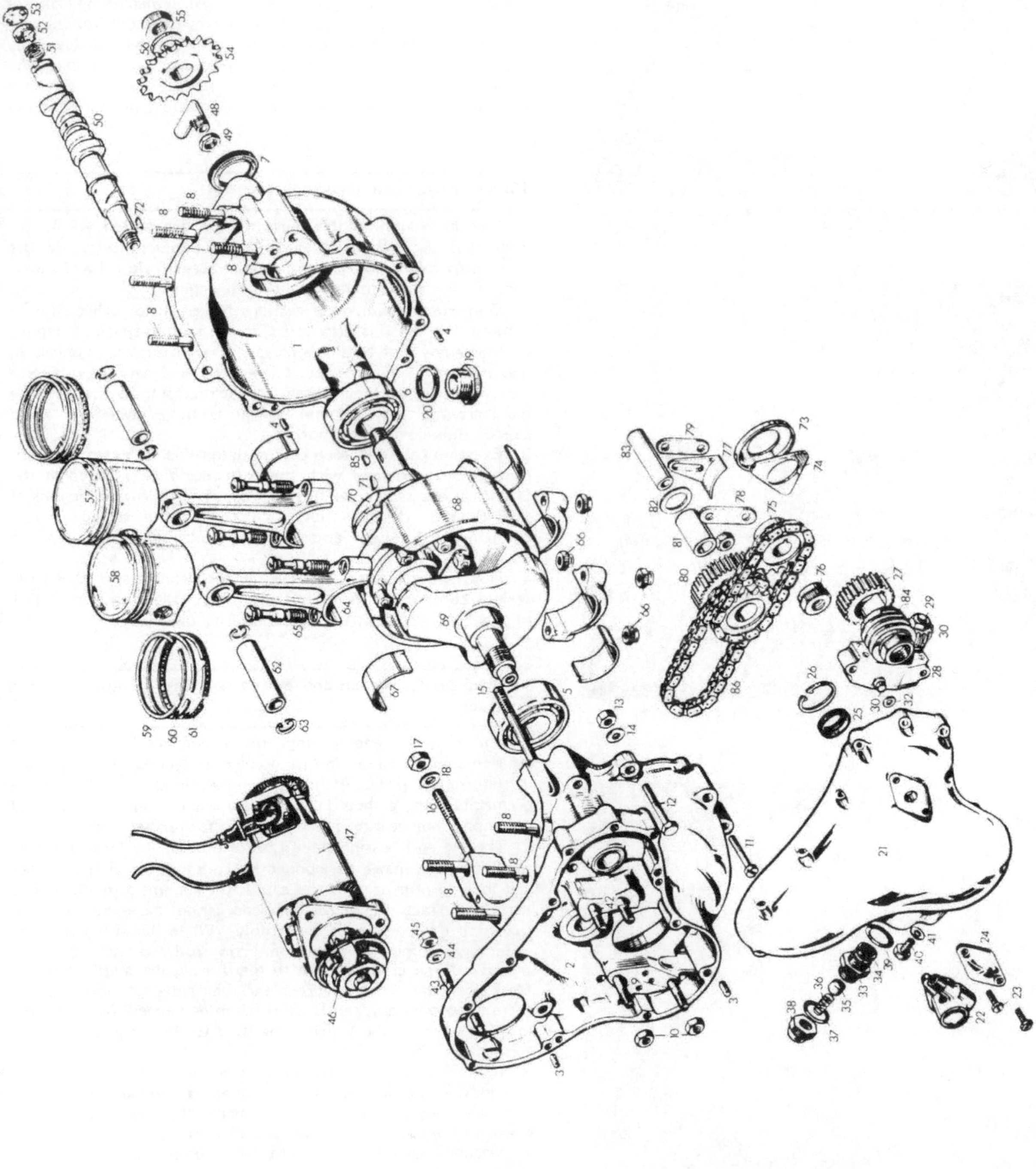

FIG. 1.6 CRANKCASE ASSEMBLY (750 cc type)

1 Crankcase, driving side only
2 " timing side only
3 " dowel - 2 off
4 " dowel - 2 off
5 Mainshaft bearing, timing side
6 Mainshaft roller bearing, driving side
7 Driving side shaft oil seal
8 Cylinder barrel studs - 9 off various diameters
9 Oil pump stud - 2 off
10 Oil pump stud nut - 2 off
11 Screw, Timing to driving side - crankcase sump (2 off)
12 Crankcase bolt (short)
13 Nut for item 12
14 Washer for item 12 - 2 off
15 Crankcase top stud (Front)
16 Crankcase top stud (Rear)
17 Nut for item 16
18 Washer for item 16
19 Crankcase oil sump filter body
20 Washer for item 19
21 Timing Cover (for tachometer fixing)
22 Tachometer gearbox
23 Mounting screw for item 22 - 2 off
24 Blanking plate for tacho drive
25 Mainshaft oil seal in timing cover
26 Circlip for item 25
27 Half time pinion
28 Oil pump assembly
29 Nut for oil pump spindle
30 Worm gear wheel on pump
31 Feed bush for pump
32 Sealing washer for feed bush

33 Pressure release body only
34 Wire gauze complete
35 Pressure release piston
36 " " spring
37 " " union nut washer
38 Pressure release body nut
39 Washer for pressure release union
40 Setscrew for oilway blank
41 Washer for pin
42 Stud for chain tensioner 2 off
43 Magneto mounting stud 2 off
44 Washer for item 43 3 off
45 Nut for item 43 - 3 off
46 Advance and retard unit
47 Magneto (or distributor unit)
48 Elbow for crankcase breather, 750cc only
49 Nut for item 48
50 Camshaft
51 Spring for breather
52 Rotary plate for timing breather
53 Stationary plate camshaft breather
54 Engine sprocket
55 Retaining nut for rotor
56 Washer for item 55
57 Piston - right-hand
58 Piston - left-hand
59 Top compression ring
60 2nd compression ring
61 Oil scraper ring assembly
62 Gudgeon pin - 2 off
63 Circlip - 4 off
64 Connecting rod - 2 off
65 Connecting rod bolt - 4 off

66 Connecting rod nut - 4 off
67 Big-end shell bearing - 4 off
68 Centre flywheel
69 Left-hand crank cheek
70 Right-hand crank cheek
71 Woodruff key
72 Woodruff key
73 Oil thrower
74 Triangular washer
75 Camshaft sprocket
76 Camshaft sprocket retaining nut
77 Camshaft chain tensioner
78 Outer tensioner plate
79 Inner tensioner plate
80 Intermediate pinion and sprocket
81 Intermediate pinion bush
82 Thrust washer
83 Intermediate pinion spindle
84 Oil pump drive worm
85 Woodruff key
86 Camshaft timing chain

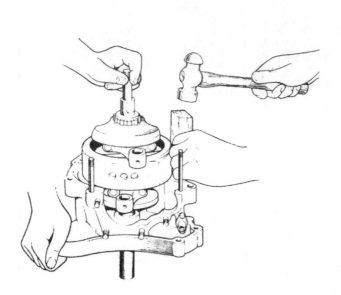

Fig. 1.7. Supporting the crankcase whilst removing the crankshaft assembly

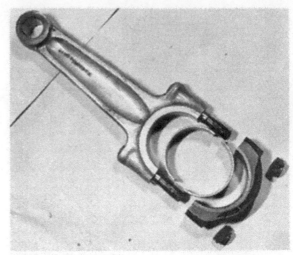

14.2 A con-rod assembly

14.4 Slacken nuts on right-hand side of assembly

14.5 Before crankshaft assembly is separated, mark flywheel to prevent accidental reversal

flanges from the flywheel with a hammer and a soft metal drift. This will release the left hand (drive side) crankshaft, which will come away with the lower two studs, tab washer and nuts. If the machine has covered a considerable mileage, it will probably be found that there is a build up of sludge in both crank flanges and in the flywheel recess as a result of the centrifugal action of the rotating assembly. This must be cleaned out thoroughly prior to reassembly.

6 Wash each crank in turn with clean petrol and use compressed air to dry off. Light score marks on the big end journals can be removed by the use of fine emery cloth but if the scoring is excessive or deep, or if measurements show ovality of more that 0.0015" to 0.002" below the low limit, the journals must be reground. The accompanying illustration shows the regrind sizes permissible; shell bearings are available in undersizes to match, from minus 0.010 inch to 0.040 inch, in 0.010 inch stages. The big end shells are finished to give the required diametrical clearance and must not be scraped to improve the fit

7 The main bearings are a tight fit in the crankcase halves - a ball journal bearing in the timing side and the outer race of a roller bearing in the drive side. They should only be removed after the crankcase has been heated in an oven; with the casing hot, drop the inside face onto a wooden bench to shock the bearings out of position.

8 Main bearing failure is characterised by a rumbling noise from the engine and some vibration. Bearings of the ball or roller type should be renewed if any play is evident, if the tracks are worn or pitted, or if any roughness is felt when they are rotated by hand.

9 The drive side main bearing has an oil seal which fits into a recess in the outer side of the crankcase. This seal should be removed and renewed. Take care not to damage the seating area in the crankcase when driving it out.

FIG. 1.8 CRANKSHAFT REGRINDING DATA

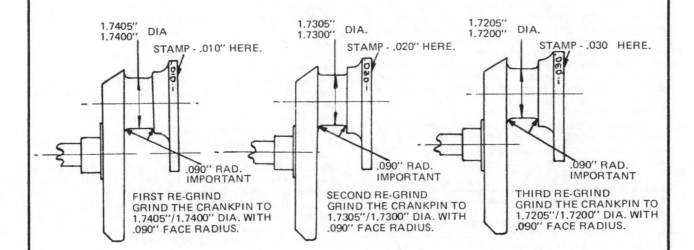

FIG. 1.8a 650 and 750cc TWINS

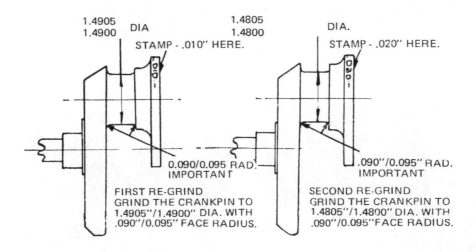

FIG. 1.8b 500cc and 600cc TWINS

15 Timing pinions, timing chain and chain tensioner - examination and renovation

1 It is unlikely that either the timing pinions or sprockets will require attention unless a timing chain breakage has resulted in either chipped or broken teeth. These components have an exceptionally long life and rarely need renewing. They will, however, require attention if main bearing failure has occurred.

2 The timing chain is of the endless type and should be examined closely for any signs of broken rollers or cracked side plates. If the chain has worn unevenly, making it difficult to adjust the chain tensioner, this is another cause for renewal. If there is any doubt about the condition of the chain it is wise to renew it, since a breakage may damage the timing pinions and sprockets, bending the valves and pushrods.

3 Although the chain tensioner will be grooved from sliding contact with the side plates of the timing chain, there is no necessity for renewal unless the grooves are deep and there is no further adjustment left.

16 Timing cover oil seal - examination and renewal

1 The oil seal which is fitted in the timing cover forms a link in the pressure feed of oil to the big ends. Consequently, the condition of this seal is very important to the life of the engine and it should be renewed every time the cover is removed.

2 Remove the seal by first extracting the retaining circlip then lever through the centre of the seal with a screwdriver, from positions 120° apart, ensuring that the mating surface of the seal in the timing cover is not scratched in any way.

3 Carefully press or drift the replacement seal into position such that the metal covered face is towards the crankshaft when assembled. Replace the circlip.

4 There is also a conical shaped seal which surrounds the high pressure exit nipple of the oil pump. This seal is operational under pressure from the timing cover and must also be renewed whenever the cover is removed. On reassembly, check that the free gap between the cover and the crankcase face is 0.010in. If the gap is less, or the seal mutilated in any way, the seal must be renewed and packed out with spacers (if required) to obtain the 0.010in gap.

17 Cylinder barrel - examination and renovation

1 The usual indications of badly worn cylinder bores and pistons are excessive oil consumption and piston slap, a metallic rattle which occurs when there is little or no load on the engine. If the top of the cylinder barrel is examined carefully, it will be observed that there is a ridge on the thrust side of each cylinder bore which marks the limit of travel of the uppermost piston ring. The depth of this ridge will vary according to the amount of wear that has taken place and can therefore be used as a guide to bore wear.

2 Measure the bore diameter below each ridge, using an internal micrometer. Compare this reading with the diameter at the bottom of each bore. If the difference in readings exceeds 0.005 inch (0.1270 mm) it is necessary to have the cylinder barrel rebored and to fit oversize pistons and rings.

3 If an internal micrometer is not available, the amount of wear can be measured by inserting a compression ring so that it is about ½ inch from the top of the bore and seated squarely in the bore by pressing it down with the skirt of the piston. Measure the ring gap with a feeler gauge, then reposition the ring below the area traversed by the piston and measure the gap again. Subtract the second reading from the first and divide the difference by three to give the diametrical wear. If in excess of 0.005 inch (0.1270mm) a rebore is necessary.

4 Check the surfaces of both cylinder bores to ensure there are no score marks or other signs of damage that may have resulted from an earlier engine seizure or displacement of one of the circlips. Even if the bore wear is not sufficient to necessitate a rebore, a deep indentation will override this decision in view of the compression leak that will occur.

5 Check that the external cooling fins are not clogged with road dirt or oil, otherwise the engine may overheat. Clean off the barrels and take the opportunity to apply a coat of matt black paint to help cooling.

6 Examine the base flange of the cylinder barrel. If the engine has been overstressed by excessive tuning, the flange is one of the first parts to fail, usually around the root of the bores. If the flange is cracked, renewal of the cylinder barrel is essential.

7 The cam followers are located in the base of the cylinder block, where each pair are retained by a locating plate and two

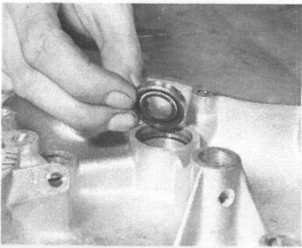

16.3a Crankshaft oil seal is driven into position, then ...

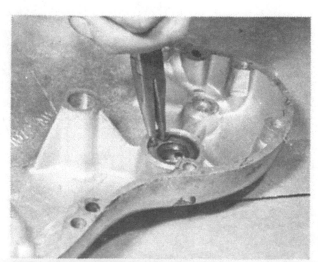

16.3b ... then retained in position by a circlip

setscrews, wired together. Unless the hardened surface is badly worn, chipped or has broken through, there is no necessity to remove them. Note, however, that when the camshaft is renewed, the followers must also be renewed as a set irrespective of their condition, otherwise premature wear of the camshaft will take place, with excessive mechanical noise.

18 Pistons, piston rings and small ends - examination and renovation

1   If a rebore is necessary, the pistons and rings can be discarded because they must be replaced by their oversize counterparts.

2   Remove all traces of carbon from the piston crowns, using a soft scraper to ensure the surface is not marked. Finish off by polishing the crowns with metal polish, so that the carbon will not adhere so readily. NEVER use emery cloth.

3   Piston wear usually occurs at the base of the skirt and takes the form of vertical streaks or score marks on the thrust side. If a previous engine seizure has occurred, the score marks will be very obvious. Pistons which have been subjected to heavy wear or seizure should be rejected and new ones obtained.

4   The piston ring grooves may become enlarged in use, permitting the rings to have greater side float. It is unusual for this type of wear to occur on its own, but if the side float appears excessive, as characterised by brown stains on the ring, new pistons of the correct size should be fitted.

5   Piston ring wear is measured as detailed in Section 17.3. If the end gap in the two positions is near identical, but is greater than the recommended limit of 0.018 - 0.020 inch, the piston rings are worn and must be renewed.

6   The gudgeon pins must be a good sliding fit in the small end of the connecting rods without evidence of play. 650 cc and 750 cc connecting rods are not bushed and must be renewed if excessive small end wear occurs. Worn small ends produce a rattle, not unlike piston slap, which will rapidly increase in intensity.

19 Valves, valve springs and valve guides - examination and renovation

1   Before the valves, valve springs and valve guides can be examined, it is preferable to remove the rocker arms and valves from the cylinder head to give better access for a valve spring compressor. Commence by withdrawing the rocker spindles. (See Section 6.16 of this chapter).

2   Insert a valve spring compressor and release each of the valves in turn. Keep the valves, valve springs and collets etc together in sets so that they are eventually replaced in their original location.

3   After cleaning all four valves to remove carbon and burnt oil, examine the heads for signs of pitting or burning. Examine the valve seats in the cylinder head. The exhaust valves and their seats will require the most attention because they are the hotter running. If the pitting is slight, the marks can be removed by grinding the seats and valve heads together, using fine valve grinding compound.

4   Valve grinding is a simple, if somewhat laborious task. Smear a trace of fine valve grinding compound (carborundum paste) on the seat face and apply a suction grinding tool to the head of the valve. Oil the stem of the valve and insert it in the guide until it seats in the grinding compound. Using a semi-rotary motion, grind-in the valve head to its seat, using a backward and forward motion. It is advisable to lift the valve occasionally to distribute the grinding compound more evenly. Repeat this application until an unbroken ring of light grey matt finish is obtained on both valve and seat. This denotes the grinding operation is now

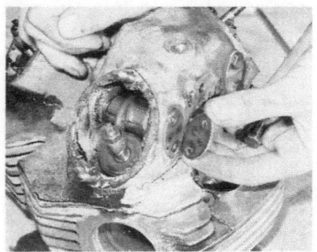

19.1a Remove rocker cover plates first

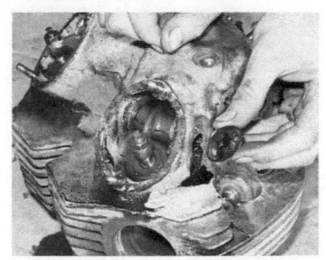

19.1b Locking plate locates with end of rocker spindle

19.2 Spring compressor can be fitted without removing rocker spindles

complete. Before passing to the next valve, make sure that all traces of the valve grinding compound have been removed from both the valve and its seat and that none has entered the valve guide. If this precaution is not observed, rapid wear will take place due to the highly abrasive nature of the carborundum base.

5 When deep pits are encountered, it will be necessary to use a valve refacing machine and a valve seat cutter, set to an angle of 45°. Never resort to excessive grinding because this will only pocket the valves in the head and lead to reduced engine efficiency. If there is any doubt about the condition of a valve, fit a new one.

6 Examine the condition of the valve collets and the groove on the valve stem in which they seat. If there is any sign of damage,new parts should be fitted. Check that the valve spring collar is not cracked. if the collets work loose or the collar splits whilst the engine is running a valve could drop in and cause extensive damage.

7 Measure the valve stems for wear, comparing them with the unworn portion that does not extend into the valve guide. Check also the valve guides for excessive play. Valve stem diameter, when new, is 0.310 - 0.3115 ins. Check that the end of the stem is not indented from contact with the rocker arm, making tappet adjustment difficult.

8 Check the free length of each valve spring and replace the whole set if any has taken a permanent set. The free length is as follows:

Outer springs 1.700ins } Post-1959 engines
Inner springs 1.531 ins }

Worn or 'tired' valve springs have a marked effect on engine performance and should preferably be renewed during each decoke as a minimum, especially in view of their low overall cost.

9 The cast iron valve guides are a tight interference fit in the aluminium alloy cylinder head and can be removed and refitted only after the cylinder head has been heated to a temperature in the region of 150° - 200°C.

The cylinder head MUST be heated to the CORRECT TEMPERATURE during this operation.

10 Oversize valve guides in plus 0.002 inch, 0.05 inch, 0.0010 inch and 0.015 inch sizes are available as replacements. The valve guide to cylinder head interference should be within the range 0.0015 - 0.0025 inch. If there is ovality, necessitating the use of oversize valve guides, the bores in the head must be reamed oversize to suit. When new valve guides are fitted, it will be necessary to recut the valve seats unless the correct Norton tool has been used to renew the valve guides.

11 In order to assess the wear on the valve guides, remove the valve springs and allow the valve to drop slightly into the combustion space. Move the valve backwards and forwards and then side to side. If the former movement is very much larger than the latter, the guide is in need of replacement.

20 Cylinder head - examination and renovation

1 Remove all traces of carbon from the combustion chambers and the inlet and exhaust ports, using a soft scraper which will not damage the surface of the valve seats. Finish by polishing the combustion chambers and ports with metal polish so that carbon does not adhere so readily. Never use emery cloth since the particles of abrasive will become embedded in the soft metal.

2 Check to make sure the valve guides are free from oil sludge or other foreign matter that may cause the valves to stick.

3 If the valve seats are pocketed, as the result of excessive valve grinding in the past, the valves seats should be recut. However, if recutting produces little "meat" around the part the valve seats should be re-inserted.
This is a specialist task which requires expert attention and is quite beyond the means of the average owner.

4 Make sure the cylinder head fins are not clogged with oil or road dirt, otherwise the engine may overheat. If necessary, use a

wire brush but take care not to damage the light alloy fins which, in places, are thin in section.

21 Rocker arms and rocker spindles - examination and renovation

1 Examine carefully the outer surfaces of each rocker arm, to ensure there are no surface cracks or other signs of premature failure. The rocker arms should have a smooth surface, to resist any tendency towards fatigue failure.

2 The rocker arms should be a good sliding fit on the rocker spindles without excessive play. Noisy valve gear will result from worn rocker arms and spindles and performance may drop off as a result of reduced valve lift. If play is evident, the rocker arms should be renewed and new spindles fitted.

3 Check the rocker arm adjuster and ball end, and the ball end of the rocker which engages with the pushrod. Both these points of contact have hardened ends and it is important that the surface is not scuffed, chipped or broken, otherwise rapid wear will occur. The rocker adjuster and locknut can be renewed as separate items.

4 The rocker spindles must have a smooth,polished surface and an unobstructed oil way. Wear is most likely to occur if the flow of oil to the rocker gear is impeded in any way. For example, if the external rocker feed pipe unions are tightened carelessly, it is possible to 'neck' the thin pipe near the unions as the result of twisting action and seriously reduce the rate of oil flow.

22 Camshaft and pushrods - examination and renovation

1 The camshaft is unlikely to show signs of wear unless a high mileage has been covered or there has been a breakdown in the lubrication system. Wear will be most obvious on the flanks of the cams and at the peak, where flattening-off may occur. Scuffing, or in an extreme case, discoloration, is usually indicative of lubrication breakdown.

2 If there is any doubt about the condition of the camshaft, it is advisable to renew it whilst the engine is completely dismantled. Comparison with a new camshaft is often the best means of checking visually the extent of wear.

3 Check the pushrods for straightness by rolling them on a flat surface. Renew any that are bent, since it is impractical to straighten them with accuracy. Check that the hardened end pieces are not loose, or the internal bearing surfaces worn, chipped or broken.

4 Unless the machine is to be used for racing, no advantage is gained by fitting a 'Domi Racer' camshaft. However early machines may gain a little extra performance by fitting the camshaft of the 650cc series (as used on the S.S.and ATLAS models), along with the push rods and valve springs of the S.S. models. Note: Sports camshafts produce an increase in mechanical noise and wear of the valve gear.

23 Engine reassembly - general

1 Before reassembly, the various engine components should be thoroughly clean and laid out close to the working area

2 Make sure all traces of old gaskets and gasket cement have been removed and that the mating surfaces are clean and un-damaged. One of the best ways to remove old gasket cement is to apply a rag soaked in methylated spirits. This acts as a solvent and will ensure the cement is removed without resort to scraping, with the consequent risk of damage.

3 Gather together all the necessary tools and have available an oil can filled with clean engine oil. Make sure all the new gaskets and oil seals are to hand, also any replacement parts required. There is nothing more infuriating than having to stop in the middle of a reassembly sequence because a vital gasket or replacement part has been overlooked.

4 Make sure the reassembly area is clean and well lit and that there is adequate working space. Refer to the torque and clearance settings, wherever they are given. Many of the smaller bolts are easily sheared if they are overtightened. Always use the correct size spanner and screwdriver, never an adjustable or grips as a subsitute. If some of the nuts and bolts that have to be replaced were damaged during the dismantling operation, renew them. This will make any subsequent reassembly and dismantling much easier.

5 Above all else, use good quality tools and work at a steady pace, taking care that no part of a reassembly sequence is omitted. Short cuts invariably give rise to problems, some of which may not be apparent until a much later stage.

24 Engine reassembly - rebuilding the crankshaft assembly

1 Before the crankshaft is reassembled, check that all parts are clean and that the oilways are free, preferably by blowing through with compressed air. Arrange the parts in the correct order for reassembly, taking note of the alignment marks made when the crankshaft was dismantled.

2 Fit the left hand (drive side) crankshaft to the centre flywheel so that the ends of the two lower studs will locate with the right hand (timing side) crankshaft when both are mated up with the flywheel. A dowel in each face of the flywheel aids location.

3 Insert the four remaining studs from the left-hand (drive) side and fit all twelve nuts. Note that the nuts fitted to the two lower studs should have a locking tab on each side of the flywheel, which must be bent over when the nuts are tightened fully. The four studs may need to be tapped into position with a drift because they are a good fit in the crank flanges. Tighten all six nuts in a diagonal sequence, to a torque wrench setting of 25 lb ft. Check to ensure the oilways blanking plug is fitted in the right hand (timing side) crank cheek. This is important if a new right hand (timing side) crank cheek has been fitted. Centre-punch the nuts as a safeguard against slackening.

4 Pump oil through the crankshaft assembly with a pressure oil can to ensure all the oilways are clear and that the oil flow is not impeded in any way. A considerable amount of oil will be needed before it exudes from the oilways, due to the need to fill the area within the flywheel centre and both crank flanges.

25 Engine reassembly - refitting the connecting rods

1 Arrange the connecting rods, their end caps and their bearing shells in the correct order for reassembly, noting that these parts were marked, when dismantled, to ensure they are replaced in their original locations. The connecting rods MUST be fitted with the oilways from the big end bearing facing OUTWARDS (late 650cc and 750cc models after engine no. 116372).

2 Fit the shell bearing with the central oilways into the big end eye of each connecting rod so that the locating tab aligns with the depression in the connecting rod itself. Then fit the plain shell into each end cap, again locating the shell tab in similar fashion. Fit the connecting rods after oiling both bearing surfaces and the end caps, checking that the locating marks align and that the big end oilways in the connecting rods face OUT-WARDS. Fit new connecting rod nuts and tighten them by hand. Note that dural end caps have washers under the nuts; steel end caps have no washers. Check that the big end bearings are still free, then tighten the nuts with a torque wrench to a setting of 25 lb ft. (15 lb ft on 500 cc and 600 cc models).

Check again that the big end bearings are free from binding. Note: Genuine 650 cc and 750 cc shells have an oilway drilled in one of each pair. They are not necessary on other models.

3 It is necessary to renew only the connecting rod nuts because it is the thread within the nuts which shows a tendancy to stretch after a previous tightening and not the thread of the bolts. However ensure that the new bolts are finger tight on the threads ie. they are a good fit and do not spin on the threads.

26 Engine reassembly - reassembling the crankcases

1 Support the right hand crankcase open-end uppermost on the workbench and lower the crankshaft assembly into position. Check that the right hand connecting rod is centrally disposed so that it will enter the crankcase mouth. A few taps with a rawhide mallet may be necessary, to ensure the right hand main bearing locates correctly.

2 When the crankshaft is fully home in the right hand crankcase, raise the assembly into the horizontal position and fit the oil sealing disc over the right hand end of the crankshaft, lipped side outwards. This should abut against the outer surface of the

24.2 Locating dowels aid crankshaft reassembly

24.3 Tighten to 25 ft lb setting

right hand main bearing. Fit the triangular-shaped washer, fit the Woodruff key into the crankshaft, then replace the crankshaft pinion so that the timing mark and chamfered teeth face outwards. It may be necessary to drive this pinion into position, using a tubular drift.

3 Lay the right hand crankcase on the workbench so that the crankshaft is again in a vertical position and fit the camshaft into the camshaft bush.

4 Lightly smear the mating surfaces of both crankcases with gasket cement. Insert the rotary breather disc and spring into the camshaft bush of the left hand crankcase, with the driving dogs facing inwards. Fit the camshaft so that it locates with the driving dogs and hold it under tension whilst assembling the cases. Check that the left hand connecting rod is correctly disposed in relation to the crankcase mouth, and lower the left hand crankcase into position, maintaining the breather disc and spring in position with one finger, or with a light smear of grease. Before the crankcases are finally mated together, it is necessary to check that the driving dogs of the breather disc have located correctly with the end of the crankshaft. If they have not, the assembly will jam and it will not be possible to make the crankcases meet. It may be necessary to turn the camshaft slightly during the latter stages of the reassembly operation, to facilitate alignment, and to tap the left hand crankcase with a rawhide mallet to ensure the left hand main bearing locates correctly.

5 Fit the front and rear screws which secure both crankcases together, also the short studs, nuts and washers. Check that both the crankshaft and the camshaft revolve freely without excessive end float, then fit the remainder of the crankcase retaining nuts and bolts. Tighten the various nuts, bolts and screws evenly, in rotation, and again check that the crankshaft and camshaft revolve freely.

6 Fit a new oil seal in front of the left hand main bearing (drive side). It is a light drive-in fit.

27 Engine reassembly - completing reassembly of the timing side

1 Before further reassembly, it is advantageous to return the crankcase to its bench mounted stand or refit the engine plates and gearbox to provide a stable working platform.

2 Offer up the engine plates and slide in, from the left, the 3 engine bolts and split washers. Lightly attach the right hand nuts and washers; slide in the footrest spacer and locate it with the footrest bar.

3 Position the gearbox and replace the upper nut and bolt, noting that the head of the bolt has flats which mate with the left hand engine plate. Slide in the lower stud from the right remembering that the left hand nut need not be replaced at this point because it locates the primary chaincase.

4 Tighten the 3 engine bolts and the upper gearbox bolt.

5 Turn the crankshaft assembly so that the timing mark on the crankshaft pinion is uppermost (12 o'clock position). Loosely position the camshaft tensioner over its retaining studs, noting that the thinner of the two clamping plates slides over the studs first, followed by the tensioner arm, then the thicker clamping plate. Fit, but do not tighten the nuts and star washers.

6 Check that the intermediate pinion/sprocket will slide onto its shaft without fouling the tensioner assembly. If not, remove the tensioner plates and file a recess to suit with a round file. Replace the tensioner and the intermediate pinion thrust washer.

7 Assemble the intermediate gear pinion with integral sprocket, the camshaft sprocket and the camshaft chain. Both sprockets have a timing mark and must be positioned so that they are ten chain rollers apart. (See fig. 1.9). The gear pinion behind the chain sprocket has a timing mark, which must register exactly with the mark on the crankshaft pinion, with which it engages. A paint mark on the intermediate gear pinion in the vicinity of the timing mark aids location. When the timing marks register, slide the two camshaft chain sprockets onto their respective shafts,

25.1 Connecting rod oilway MUST face outwards - late 650cc and 750cc engines

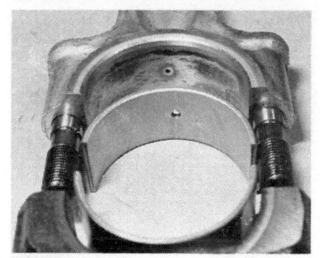

25.2a Ensure oilways align on shell and rod on 650 cc and 750 cc engines.

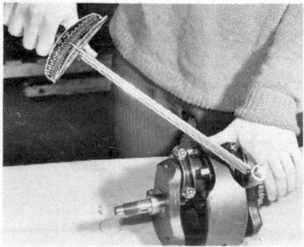

25.2b Torque-up the big end bolts

after inserting the Woodruff key into the camshaft to ensure correct register with the camshaft sprocket. It may be necessary to tap this latter sprocket into position. When the engine is at top dead centre the timing marks on the sprockets should be in the 11 o'clock position.

8 Before continuing with the reassembly, again check the accuracy of the timing, to ensure all the timing marks are in correct register. It is easy to rectify any error at this stage.

9 After positioning the rocker feed pipe on later engines, replace the magneto or distributor, as fitted, by reversing the stripdown procedure in section 10. On early models the dynamo should also be replaced at this stage. The ignition timing is covered in Chapter five and should be completed before the timing cover is replaced.

10 Adjust the chain tensioner to give a maximum of 1/8 inch play at the tightest point of the chain run. Check the tension with the crankshaft in several different positions before tightening the chain tensioner nuts fully, to obviate the possibility of tight spots giving a false reading.

11 Fit the oil pump worm on the end of the crankshaft. The nut has a LEFT HAND thread. It can be tightened fully by locking the engine with a metal rod passed through the small eyes of both connecting rods. Use Loctite to retain it firmly.

12 Whilst the engine is still locked, fit the camshaft nut. This has a normal right hand thread; also, note that this nut drives the tachometer gearbox (where fitted). Use Loctite on this nut too.

13 Fit the oil pump. No gasket or cement is used on this joint on the Dominator series. However, a visual check on the crankcase/oil pump seal should be made: a "Commando" gasket is available for this joint if required. Check to ensure the gasket aligns correctly with the oilways; if it is accidentally reversed, they may be masked off. The pump is secured by two nuts without washers, which must be tightened to a torque setting of 10 to 12 lb ft. If a thick timing cover gasket is to be used, fit an oil pump gasket too. Fit it to the pump and NOT the crankcase.

14 If the oil pump has been dismantled, it must be primed with oil after it has been fitted. Turn the engine with an oil gun pushed up to the oil feed hole in the crankcase rear.

15 Fit a new conical rubber seal over the oil pump outlet and offer up the timing cover. The oil seal should push the cover away from the crankcase forming a gap of 0.010 in. If the gap does not exist, the seal should be packed out with shims between the seal and the pump body until the gap is obtained - do not over shim because if the seal is over-compressed by the timing cover, permanent distortion will ccur, with risk of leakage.

16 Check the tightness of the nut securing the oil pump driving gear.

17 Before refitting the timing cover, check that the mating surfaces are clean and undamaged and that the oil seal within the inner face of the cover is in good condition; it is a push fit and retained by a circlip. See Section 16 of this Chapter.

18 The timing cover has a pressure relief valve fitted at the rear- under the domed nut. This may be removed for cleaning: DO NOT alter the washers fitted as they control the length of the spring determining the operating pressure of the relief valve.

19 Before fitting the timing cover, check that the blanking plug fitted below the body of the pressure release valve is located correctly. The rocker oil feed bolt fits here on later models. Fill the crankshaft with oil here.

20 Lightly smear both mating surfaces with gasket cement, and use a new jointing gasket. Refit the timing cover, noting that the retaining screws are positioned according to their length, as shown in the accompanying illustration. Fig 1.10. Tighten them fully. Gasket cement is not needed for the later thick gasket.

---

28 Engine reassembly - refitting the pistons and piston rings

1 Assemble the piston rings on each piston.
2 On the 750 models. and some 650 models, the oil control

26.1 Support timing side crankcase above bench during reassembly

26.2a Fit oil sealing disc and triangular washer,...

26.2b ... followed by Woodruff key and crankshaft pinion

26.3 Insert camshaft in front of crankcase

26.4a The crankcase breather assembly

26.4b Lower drive side crankcase into position

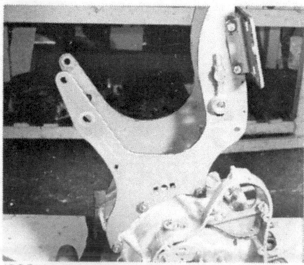
27.2 Engine plates attach to crankcase by 3 bolts

27.3 Offer up the gearbox for attachment

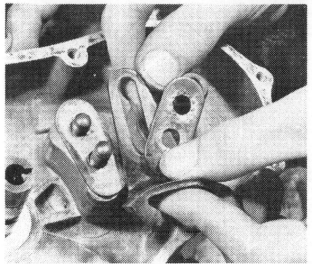

27.5 The tensioner assembler

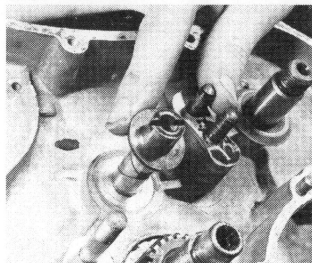

27.6 Replace the intermediate pinion thrust washer

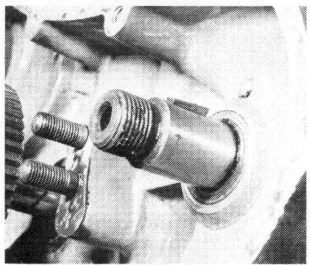

27.7a Replace the Woodruff key in the camshaft

27.7b Slide the camshaft drive tensioner into position

27.10 Maximum play in chain should be 1/8 in

27.13 Refit the oil pump. Note: conical oil seal washer

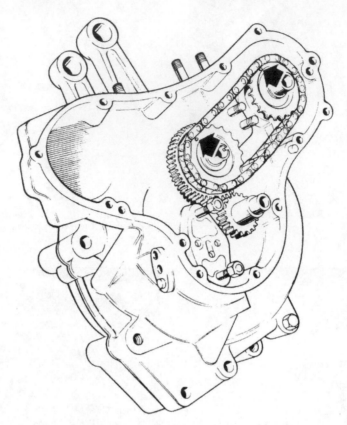

**Fig. 1.9. Alignment of timing marks for correct valve timing.
Marks must be ten chain rollers apart**

rings are built up from three or even five component parts, an
expander and top and bottom rails. Ensure that the ends of the
expander meet each other without overlapping. Current issues
of this ring assembly have the ends painted green and white -
these two colours should be seen when the expander is in
position. If only one colour can be seen, then the expander is
incorrectly fitted (see Fig. 1.11).

3   Unless new rings are being fitted, it is not advisable to remove
the carbon from the bottom of the ring groove on the back of
the ring. Even so, it is important to ensure that the rings are free
to move and not 'gummed' in position.

4   Warm the pistons to aid insertion of the gudgeon pin and
ensure one circlip is replaced in each piston boss, sharp edge
outwards, (later type pistons).

The pistons are marked on the crown to ensure replacement
in the correct position; the inlet valve cutaway is the one farthest
away from the edge of the crown (750 cc models)

5   Oil the small end eyes of the connecting rods and the
gudgeon pins and gudgeon pin bosses. It is advisable to pad the
mouth of the crankcase at this stage with clean rag, to prevent a
misplaced circlip from falling in.

6   Support the piston and connection rod, and press each
gudgeon pin into position. Fit the second circlip in each case
with the sharp edge outwards. Make quite sure that BOTH
circlips are correctly located with the groove inside each piston
bore. A misplaced circlip will cause serious engine damage.

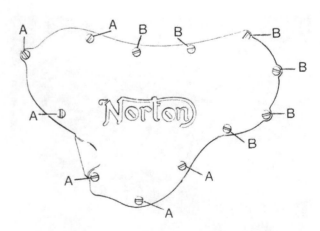

Fig. 1.10 Timing cover screws location

A   Long screw 1¼ inch
B   Short screw 1 inch

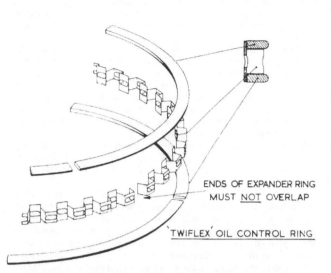

Fig. 1.11. Twinflex oil control ring construction

ENDS OF EXPANDER RING MUST NOT OVERLAP

'TWIFLEX' OIL CONTROL RING

28.5 Pad crankcase mouth during piston replacement

29 Replacing the crankcase assembly in the frame

1 It is convenient to replace the crankcase assembly in the frame at this stage, whilst the engine unit is still comparatively light in weight. Lift the engine/gearbox unit into the frame from the right hand side: the kickstarter and the timing cover should be temporarily fitted for this operation.

2 Allow the crankcase to rest on the bottom frame tubes and replace the two bolts which retain the engine plates to the upper frame cross-members at the rear of the gearbox. The back bolt has a plain washer under its head and a split washer under the nut

3 Lever up the engine unit with a length of wood and replace the front engine plates. Attach the plates to the engine by inserting the bolts from the left hand side. Lower the engine unit and attach the plates to the frame.

4 Replace the two, lower, rear engine plate bolts; replace the front engine plate cover (if fitted) and tighten all 8 sets of nuts and bolts.

5 The centre-stand may now be replaced. Lever the spring onto its support inside the left hand engine plate. Note. After many years, the 'eyes' of the stand and the frame support tend to wear extensively, as do the bottom of the frame tubes where the stand abuts and the retaining bolts. If wear is extensive, bushing of the eyes/welding of the frame/replacement of the bolts should be considered as appropriate. Don't forget the bolts through the stand bolts to lock them in position.

30 Engine reassembly - refitting the cylinder barrel

1 Position the piston rings so that their end gaps do not coincide and so that the end gap of the oil control ring is clear of the cutaway in the base of each cylinder bore (where applicable). On pistons with an expander ring fitted, the rail gaps should be at least one inch to either side of the expander gap, when the ring is built up. If not correctly positioned, the rail ends may become trapped with the cylinder base cutaway and break up.

2 Smear both pistons and the cylinder bores with clean engine oil, then fit a pair of piston ring compressors to each piston, if available. Fit a new cylinder base gasket (no gasket cement), such that the oil hole in the gasket aligns with the hole in the crankcase lip. Place a pair of wooden or metal rods below each piston, across the crankcase mouth, to steady the pistons as the cylinder barrel is lowered into position.

3 When the piston rings have engaged with the bores, remove the clamps and the rods below the pistons. At the same time,

withdraw the rag used for padding the crankcase mouth.

4 Before the cylinder barrel is lowered fully, engage the cylinder base nuts, since there is insufficient clearance for them to be fitted when the barrel is seated on the crankcase mouth.

5 Tighten the cylinder base nuts in sequence. The 3/8 inch nuts and bolts should be tightened to a torque setting of 25 lb ft and the 5/16 inch nuts to 20 lb ft.

31 Engine reassembly - reassembling the cylinder head

1 Replace the valves and valve springs, together with their associated seatings and collars etc.

2 If the rocker spindles have to be replaced, heat the cylinder head to a maximum of 150° - 200°C in order that the spindles may be driven back into position. Note: On the scramble framed models, only the inlet rockers should be replaced before refitting the cylinder head. Each rocker arm should have a double spring washer fitted against the innermost boss and a plain washer against the other boss, to take up end float. The outer end of the rocker spindle must be aligned so that the two slots are in a horizontal position before it is driven flush with the jointing face. The flat on the spindle must face the rocker cover in each case.

3 Oil the rocker spindles and rocker arms after reassembly and check that all four rocker arms move freely, without binding. Refit the end covers which are secured by two bolts. There is no necessity to renew the gaskets, unless there has been a previous leakage, or unless the locking plate and retaining plate have been separated.

32 Engine reassembly - reassembling the primary transmissions

Note: The following assembly procedure is for the pressed steel chaincase featherbed frame models. However, the assembly of the aluminium chaincase models is very similar and the following should be used as a guide remembering that the outer chaincase is retained by 14 screws.

1 Replace the final drive sprocket on the splines of the gearbox mainshaft - dished face outwards. Fit the retaining nut, which has a LEFT HAND THREAD, and tighten it to a torque setting of 80lb ft. Fit the locking plate and screw after slightly slackening the nut (if necessary) such that the retaining screw for the locking plate will fit in either of the two tapped holes in the sprocket. Note: This is an important operation as the nut is

prone to vibrate loose, producing high chain and bearing wear and rough transmission. Replace the final drive chain.

2   Reconnect the earth lead (magneto) or the low tension wire (coil ignition) as appropriate.

3   Ensure that the long footrest bolt and the engine plate spacer tube through which it passes are in position. Replace the sleeve spacer such that the two dogs mate with the rectangular holes in the left hand engine plate.

4   Offer up the inner chaincase - it is retained by the bottom gearbox bolt at the bottom rear; a nut and spring washer on the end of the spacer from the engine plate (central) and three cheese head screws to the crankcase. A new gasket, with a smear of sealing compound on either side, should be used at the crankcase/chaincase joint to ensure oil lightness at this point.

5   Attach the stator housing to the crankcase with the three countersunk Allen screws. These screws tend to "round off" quickly as they are not large enough for the job; replace them if any trouble was encountered during their removal.

6   Replace the Woodruff key used to locate the engine sprocket on the left hand end of the crankshaft. Also drive the sprocket into position.

7   Slide the clutch centre back onto its splines and replace the retaining spring washer and nut: the shaft may be locked by engaging top gear and locking the back wheel. Torque the nut to a setting of 70 lb ft.

8   Replace the clutch outer drum, the primary chain and spring link and slide the clutch operating rod into the centre of the gearbox mainshaft.

9   Assemble the clutch plates in the order they were removed - on very early clutches with cork inserts in the clutch outer drum the first plate is a plain plate and vice versa for the later bonded plate clutches with solid plate clutch outer drums. Note that the last plate is a "half bonded" plate with the plain side facing outwards in the latter case and a plain plate in the former case.

10 Replace the clutch outer pressure plate complete with spring cups, springs and retaining nuts (slotted). The nuts should be screwed in until their heads are flush with the top of the stud. Replace the clutch cable and adjust the cable length until all the slack is just removed. Pull the clutch lever to check the pressure plate lifts evenly.

11 The alternator rotor is retained to the crankshaft with a Woodruff key. Clean the rotor to ensure the magnetism has not attracted any metallic particles and fit the rotor over the keyway, with the name facing outwards. Replace the rotor nut and shaped washer, then engage top gear and apply the rear

brake so that the engine is locked whilst the nut is tightened to a torque setting of 70-80 ft lb. Note: If a timing disc is available, the engine should be timed before replacing the stator nut, and the timingcover replaced permanently.

12 The primary chain should be tensioned at this point by adjusting the position of the gearbox using the linkage affixed to the right hand upper gearbox bolt. Loosen the large gearbox nut and draw the gearbox back until the chain is tight. Using the forward of the two small adjusting nuts, PUSH the gearbox forward until the chain has ½ in of slack in the top run and tighten up the upper and lower gearbox bolts and the small adjusting nuts. Check the tightness of the chain in 2 or 3 places - NEVER run this chain tight or the main and gearbox bearings and the chain will wear quickly. Note: Any adjustment of the gearbox position will affect the tension of the final drive chain which must be adjusted accordingly.

13 Place the stator on its three mounting studs such that the cable break-out is on the outside, towards the outer chaincase; the break-out should be positioned in the upper right hand position and the cables fed through the support plate (welded to the inside of the inner chaincase) and out through the rubber grommet in the inner chaincase.

14 Replace the plain washers and stator retaining nuts. Tighten the nuts to a torque setting of 15 lb ft. There must be a minimum air gap between the stator coil assembly and the rotor of from 0.008 - 0.010 inch and a check should be made with a feeler gauge all the way round, with the crankshaft in different positions, in case the latter is slightly bent. If the gap is reduced at any point, misalignment of the stator mounting studs should be suspected and corrected.

15 Reconnect the alternator cable snap connectors, using the colour coding noted when disconnecting.

16 Replace the large rubber band oil seal (with the small lip or protusion facing outwards) and replace the outer chaincase.

17 The chaincase is retained by a large nut on the footrest spacer. Between the nut and the casing is a large washer with a large flat rubber seal. Tighten the nut until about 2 threads are showing. In the case of the scrambles models, replace and tighten the 14 screws around the periphery of the chaincase.

18 Replace the footrest spacer on the right hand side and then replace both footrests, the rear brake lever with its locating grease nipple and the gear-change lever complete with the gear indicator.

19 Remove the oil level plug from the outer casing and fill the chaincase with the recommended quantity and grade of oil. Reconnect the back brake rod and the brake light switch.

29.2 Engine unit resting on frame and support

30.2a Pistons equipped with piston ring compressors

30.2b Using screwdrivers to steady pistons

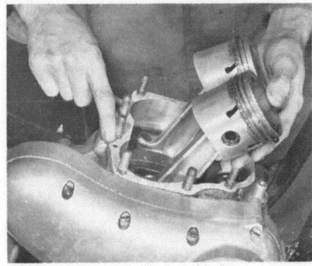

30.2c Check hole in base gasket aligns with oilway

31.1a Insert the valve ...

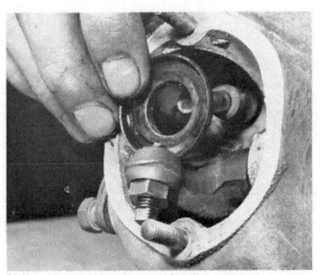

31.1b ... replace the spring seat after fitting the heat insulating washer ...

31.1c ... fit the double springs ...

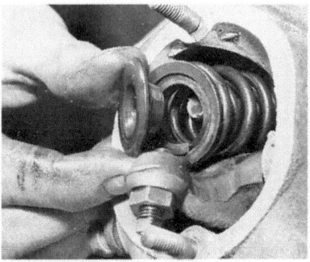

31.1d ... and the valve collar ...

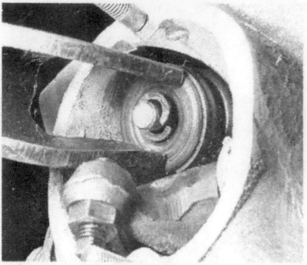

31.1e ... compress and insert collets

32.1a Final drive sprocket retained by nut - star washer removed

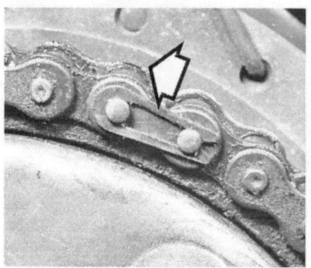

32.1b Final drive split link replaced. Arrow indicates chain travel

32.3 Sleeve spacer engages with engine plate

32.5 Countersunk Allen screws retain alternator housing

32.6 Replace the engine sprocket and key

32.7 The clutch centre is carried on splines

32.8 Slide on the clutch outer drum

32.9 The last plate is the half bonded plate: don't forget the clutch operating pushrod

32.10 Clutch pressure plate, cups, springs and nuts assembled

32.12 The primary chain adjusting linkage

32.16 Replace the outer cover. Note: position of stator lead

33 Engine reassembly - refitting the cylinder head

1   Place a new cylinder head gasket on top of the cylinder barrel and turn the engine so that the pistons are at top dead centre (TDC) to reduce valve lift to a minimum. The head gasket is asbestos sandwiched in copper.

2   Insert the pushrod in the pushrod tunnels of the cylinder head with the longer pushrods in the innermost position on each side. The cupped end of the pushrods should face upwards. Feed the pushrods into the cylinder head as far as possible.

3   Invert the cylinder head and whilst holding the pushrods with one hand, position the cylinder head over the cylinder barrel. After rotating the head through 90° to get it through the top frame tubes of the featherbed frame (If the oil tank and battery case are removed as during an engine rebuild, the head may be easily inserted from the rear of the barrels). Allow the pushrods to drop into the tunnels of the cylinder barrel, where they will locate with the cam followers.

4   On scrambles framed models, the exhaust rockers should now be replaced taking EXTREME care not to drop any of the parts down the pushrod tunnel.

5   Lower the cylinder head and engage the ends of the pushrods which are slightly higher with the ball ends of the rockers with which they engage. Some skill is necessary during this operation using a screwdriver or a piece of stout wire to guide the pushrods into position through the exhaust rocker cover orifice. When they have engaged correctly, tackle the second pair in similar fashion.

6   The short central cylinder head bolt and washer should be tightened down first to overcome the spring pressure of the valves on lift and cause the head to seat squarely on the head gasket. Tighten down the cylinder head using the sequence detailed in the accompanying illustration. See fig. 1.12. Check all four pushrods have engaged correctly before commencing the tightening sequence. The torque settings are:

3/8 inch nuts and bolts	30 lb ft
5/16 inch bolts	20 l b ft

33.1 Always use a new cylinder head gasket (not fitted to later 650 cc or very late Atlas models)

33.2a Diagrammatic view of pushrod arrangement

33.2b Pushrods installed in cylinder head

33.3 A hand is required to retain the pushrods in position

34 Engine reassembly - completion

1 Refit the carburettor(s). Do not overtighten the carburettor flange nuts as there is a risk of the flange bowing and causing air leaks. Replace the slide and needle assembly ensuring that the needle enters the needle jet and that the air slide locates with the slot on the jet block. Replace the top of each mixing chamber. Check that the twist grip and air control lever (where fitted) operate smoothly without any of the carburettor components jamming.

2 Replace the oil tank; do not forget the rubber mat upon which it sits or the rubber grommet which is part of the upper support.

3 Replace the oil breather connection and the breather/tank and tank/chainguard pipes as appropriate.

4 Reconnect the oil feed and return pipes to the timing side crankcase (one nut and washer). Use a gasket but no gasket cement. The pipe from the oil tank filter goes to the outside pipe on the crankcase.

5 Replace the battery case and reconnect any electrical connections removed ie. earth wires or rectifier connections.

6 Install the air filter (if fitted) between the battery case and oil tank and reconnect the rubber ducting.

7 Temporarily replace both spark plugs, after checking to ensure they are gapped correctly and not fouled.

8 Refit the cylinder head steady - 2 nuts, bolts and spring washers to the frame and a nut and spring washer on the stud into the head (featherbed frame models).

9 Refit the ignition coil (distributor models only) and the horn.

10 Connect the spark plug caps.

11 Replace and connect-up the battery. Fit the battery retaining clamp and the battery cover.

12 Refit the exhaust system by reversing the dismantling procedure. A new sealing washer should be fitted within each exhaust port. Ensure both finned exhaust pipe clips are tightened fully - if they work loose the internal threads of the cylinder head will be shattered away. It is best to use Norton Villiers service tool 063968 for tightening the finned rings without damage or a 'C' spanner of the appropriate diameter.

13 Reconnect the rocker feed pipe to the rear of the timing cover and to the cylinder head by means of the banjo unions provided (later models only). Use new alloy sealing washers at the union joints. Replace the pressure release valve in the rear of the timing cover, if it has been removed for cleaning, and tighten to a torque setting of 25 lb ft. Check that the crankcase drain plug filter plug has been replaced and tightened, then refill the oil tank with 4.5 Imperial pints of new engine oil.

35 Checking and resetting the valve clearances

1 In order to rotate the engine more easily, remove both spark plugs. The spark plugs were fitted at an earlier stage to preclude the possibilty of washers being dropped into the engine whilst reassembling.

2 Rotate the engine until the left hand inlet valve is open fully, then check the clearance of the right hand inlet valve. This check should be made with the engine cold. If the clearance is correct, the following feeler gauges will be a good sliding fit:-

Model 7,77,88,99 - 0.003 ins
Model 88 (SS), 99 (SS), all 650 and 750 - 0.006 ins.

3 If the clearance is not correct. slacken the locknut on the end of the rocker arm and turn the square end of the adjuster until the correct setting is achieved. Hold the square end of the adjuster steady, then tighten the locknut, then recheck the clearance. If correct proceed to the left hand inlet valve, which is checked in similar fashion whilst the right hand inlet valve is open fully.

4 Use the same technique to check and if necessary adjust the two exhaust valves. The right hand valve must be open fully when the left hand valve clearance is checked, and vice versa. Note that in the case of the exhaust valves, the correct clearance is:-

Model 7,77,88,99 - 0.005 ins
Model 88 (SS), 99 (SS), all 650 and 750 - 0.008 ins.

5 When the correct settings have been obtained for all four valves, fit new gaskets and replace the two exhaust and one inlet rocker covers. No gasket cement should be necessary if the jointing faces are in good condition and the covers tightened fully.

6 Replace both spark plugs and reconnect the plug caps.

36 Starting and running

1 Replace the petrol tank and reconnect the fuel lines. Switch on the tap(s) and check for fuel leaks.

2 Check that the electrical system is functioning correctly by means of the lights and the ammeter reading.

3 Recheck the tension of the primary and secondary chains in several places. Adjust if necessary, lock up both systems and recheck.

4 Check that both brakes work effectively and that all controls, in particular the clutch, are working correctly and freely. Refer to Chapter 37 if the clutch requires adjustment.

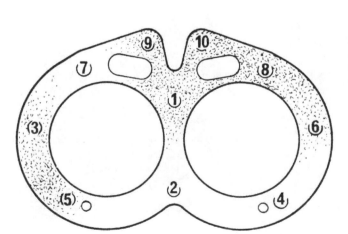

Fig. 1.12. Tightening sequence for cylinder head nuts and bolts

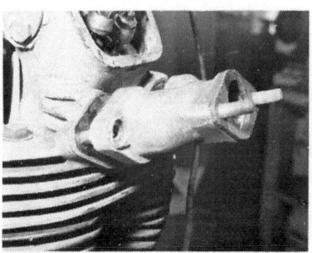

34.1a Don't forget the allen screws within this type of manifold

34.2 Do not omit the rubber mountings on the oil tank

34.8 Clyinder head steady in position

34.12 New sealing washers should be used in the exhaust ports

34.13a Screw in the rocker feed banjo bolts

34.13b Fill the oil tank with 4.5 imp pints

35.3 Set left hand inlet when right hand inlet is fully depressed

5   Switch on the ignition and start the engine. Run it at fast tick-over speed until oil commences to return to the oil tank. There may be a time lag before the flow of oil issues from the return pipe because pressure has to build up in the rebuilt engine before circulation is complete. Do not permit the engine to run at low speed for more than a couple of minutes without evidence of the oil returning, before stopping it and checking the lubrication system. To verify whether the oil pump is working, slacken the pressure relief valve a little and see whether oil emerges from around the threads, when the engine is restarted.

6   If the engine refuses to start, despite evidence of a good spark and petrol in the carburettor, try changing over the plug leads. It is easy to transpose them, especially if the engine was timed using a different procedure.

7   The exhaust will smoke excessively during the initial start, due to the presence of excess oil used during the reassembly of the various components. The smoke should clear gradually, as the engine settles down.

8   The return to the oil tank will eventually contain air bubbles because the scavenge pump will have cleared the excess oil content of the crankcase. The scavenge pump has a greater capacity than the feed pump, hence the presence of air when there is little oil to pick up.

9   Refit the seat and check the engine for leakages at gaskets and pipe unions etc. It is unlikely any will be evident if the engine has been reassembled correctly, with new gaskets and clean jointing faces.

10 If the engine has been rebored, or if a number of new parts have been fitted, a certain amount of running-in will be required.

Particular care should be taken during the first 100 miles or so, when the engine is most likely to tighten up, if it is overstressed. Commence by making maximum use of the gearbox, so that only a light loading is applied to the engine. Speeds can be worked up gradually until full performance is obtained with increasing mileage.

11 Do not tamper with the silencer or fit another design unless it is designed specifically for a Norton twin. A noisier exhaust does not necessarily mean improved performance; in a great many instances unwarranted modifications or the fitting of an unsuitable design of silencer will have an adverse effect on performance and petrol consumption and may even produce damage to the head, valves and pistons.

37 Increasing engine performance

1   Owing to the success of the "Domi racers" in the 60s and the "Commando" in the 70s, there are many parts available from the factory and private firms which will increase the performance of these engines.

2   The owner is reminded that tuning produces more torque and higher rpm from the engine and should ensure that the general state of the machine is adequate for tuning. For instance, the braking, lighting and handling of the older machines should be overhauled before any engine tuning is carried out. Also, there is no point in tuning a machine with a worn out gearbox, sprockets and chains as the extra power will produce little more than extra heat!

38 Fault diagnosis

Symptom	Cause	Remedy
Engine will not start	Earthed cut-out button	Temporarily disconnect lead and check for spark at plugs
	Reversed plug leads	Transpose leads. (Likely to occur during rebuilds.)
	Contact breaker points closed	Check and re-adjust points
	Flooded carburettor	Check whether float needle is sticking and clean
	Battery Flat (Coil ignition models only)	Recharge battery
Engine runs unevenly and misfires	Incorrect ignition timing	Check setting and adjust if necessary
	Faulty or incorrect grades of spark plug	Clean or replace plugs
	Fuel starvation	Check fuel lines and carburettor
Lack of power	Incorrect ignition timing (retarded)	Check and reset timing. Check action of automatic advance unit
	Chains too tight	Slacken off chains
Engine pinks	Incorrect ignition timing (over-advanced)	Check and reset timing.
	Compression ratio too high (tuned engines only)	Replace pistons if running on top grade fuel
Excessive mechanical noise	Worn cylinder block (piston slap)	Rebore and fit O/S pistons
	Worn small end bearing (rattle)	Replace bearings and gudgeon pins
	Worn big end bearings (knock)	Replace shell bearings and regrind crankshaft
	Worn main bearings (rattle)	Fit new bearings
	Push rods not located correctly	Locate push rods correctly
Engine overheats and fades	Lubrication failure	Check oil pump and oil pump drive. Check timing cover seals
Excessive oil consumption and exhaust smoke	Worn rings	Replace rings; rebore if necessary

Chapter 2 Gearbox

Contents

Specifications

Mainshaft diameter (Clutch End)	0.8105/0.8095 in.
Mainshaft diameter (Kickstart End)	0.6248/0.6244 in.
Mainshaft Bearing SKF-RLS5	5/8 x 9/16 x 7/16 in.
Layshaft Bearing SKF-6203	17 x 40 x 12 mm.
Layshaft Diameter (Clutch End)	0.6692/0.6687 in.
Layshaft diameter (Kickstart End)	0.6855/0.6845 in.
Sleeve gear bearing	1¼ x 2½ x 5/8 in.
Sleeve gear shaft situ (O/D)	1.2500/1.2495 in.
Sleeve gear bush (O/D)	0.9060/0.9055 in.
Sleeve gear bush - reamed in situ	0.81325/0.81200 in.
Layshaft bush - bare diameter	0.6875/0.6865 in.
Camplate plunger spring free length	1.500 in.

Primary Transmission

Engine Sprocket
General Standard	21 teeth.
Standard on G15 Mark II, G15CSR	22 teeth.
Other alternatives	17, 18, 19, 20, 23 teeth.

Gearbox Sprocket
General Standard	19 teeth.
Standard on G15 Mark II, G15C5, N15CS	17 teeth.

Primary chain
General	76 link, ½ x 0.305 in.
All "G15"s and "11"S	68 link, ½ x 0.305 in.

Gear Ratios

	General	G15CS
Top	4.53	4.96
Third	5.52	6.03
Second	7.57 (7.70)	8.40
First	11.60	12.65

Gearbox pinions - number of teeth
Layshaft 4th	18 teeth.
Layshaft 3rd	20 teeth.
Layshaft 2nd	24 teeth.
Layshaft 1st	28 teeth.
Mainshaft 4th	23 teeth.
Mainshaft 3rd	21 teeth.
Mainshaft 2nd	18 teeth.
Mainshaft 1st	14 teeth.

Rear Chain

General Standard
Early Models (Pre 1966)	97 links 5/8 x ¼ in.
Later Models (General)	97 links, 5/8 x 3/8 in.
G15CSR	98 links, 5/8 x 3/8 in.
Gearbox capacity (oil)	1.0 Imp pints SAE 90 EP Oil

1 General Description

The gearboxes fitted to all the Norton twins since 1957 are virtually identical. This is the unit designed by AMC Ltd and is of the four-speed, positive, right-hand footchange type.

The gearbox is a self contained unit and it is not necessary to dismantle or remove the engine to gain access. However, should its removal be required, then the engine and gearbox must be removed from the frame before the gearbox may be separated.

Most repair work can be accomplished with the gearbox in position, although it will be necessary to drain the oil and remove the inner and outer covers if the gear clusters, selector arms or complete require attention.

2 Dismantling the gearbox - removing the inner and outer covers

1 To remove the gear change lever, first unscrew the centre screw which retains the gear indicator in position. Then slacken the pinch bolt through the gear change lever and draw the lever off its splined shaft. The removal and replacement of the outer cover is facilitated by the gear change lever being in position (excepting where rear-set footrests are fitted). Replace the lever after inspecting the splines for damage.

2 Slacken and remove a similar pinch bolt through the base of the kickstarter and draw the kickstarter off its splined shaft. In this instance it is necessary to withdraw the pinch bolt

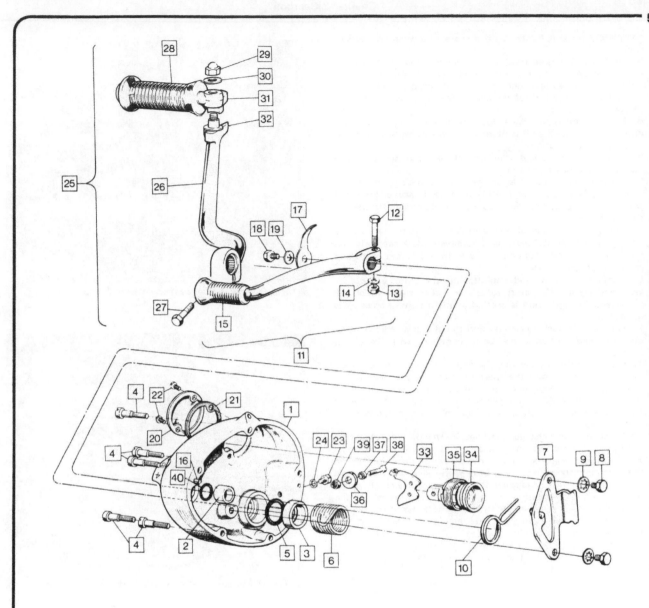

FIG. 2.1 GEARBOX OUTER COVER ASSEMBLY

1 Gearbox outer cover
2 Gearchange lever bush
3 Kickstarter lever bush
4 Outer cover screw 5 off
5 Kickstarter shaft 'O' ring
6 Kickstarter return spring
7 Gearchange stop plate
8 Gearchange stop plate bolt 2 off
9 Gearchange stop plate washer 2 off
10 Gearchange return spring

11 Gearchange lever
12 Gearchange lever bolt
13 Gearchange lever nut
14 Gearchange lever washer
15 Gearchange lever rubber
16 Dowel 2 off
17 Gearchange indicator
18 Gearchange indicator bolt
19 Gearchange indicator bolt washer
20 Inspection cover

21 Inspection cover gasket
22 Inspection cover screw 2 off
23 Oil level bolt
24 Oil level bolt washer
25 Kickstarter lever
26 Kickstarter crank only
27 Kickstarter pinch bolt
28 Kickstarter rubber
29 Domed nut
30 Plain washer

31 Pin
32 Spring washer
33 Clutch operating lever
34 Clutch operating lever body
35 Clutch operating lever body lockring
36 Clutch operating roller
37 Clutch operating roller sleeve
38 Clutch operating roller screw
39 Clutch operating roller screw nut
40 'O' ring for pawl carrier

completely since it locates with a groove between two sets of splines.

3   Undo the 2 cheese head screws which retain the inspection plate - remove the plate and its gasket. Screw in the clutch cable adjuster at the gearbox such that the clutch cable nipple may be detached from the fork of the clutch operating lever within the gearbox outer cover.

4   Place an oil drain tray below the gearbox and remove the drain plug. Allow the oil content to drain off whilst proceeding with the next operations.

5   Unscrew and remove the five cheese-head screws which secure the gearbox end cover.

6   Position the oil drain tray so that it will catch any residual oil which may be released when the outer cover is worked loose and withdrawn. The ratchet plate and spindle will remain attached to the outer cover.

7   The clutch withdrawal lever assembly is free to revolve in its housing when the locking ring is removed. Mark the body and the locking ring with punch marks so that they can be aligned correctly on reassembly.

8   Remove the screw and nut retaining the clutch withdrawal lever and detach the lever, roller and roller sleeve which are displaced. This operation is necessary only if wear or breakage is suspected.

9   Unscrew the locking ring around the clutch withdrawal lever assembly and withdraw the body complete with the clutch operating ball.

10 Select top gear by levering the end of the gear change quadrant towards the top of the aperture in the gearbox inner cover, whilst rocking the rear wheel to facilitate the engagement of the gear pinion dogs. When top gear is engaged, apply the rear brake fully and unscrew the mainshaft nut. This has a normal right hand thread.

11 Lever the gear change quadrant in the opposite direction until neutral is located, then unscrew and remove the seven nuts which retain the gearbox inner cover. Two of these nuts are on the outside of the casing, at the base of the gearbox. Remove the inner cover by lightly tapping the end of the mainshaft to aid separation, preferably after the clutch has been removed. (If the gearbox has been removed from the engine plates. If not tap the "ears" of the cover with a rawhide mallet). The cover can now be lifted away. There is no necessity to disturb the kickstarter spindle, pawl and return spring assembly unless attention to these parts is required. They will remain attached to the inner cover.

2.1 The gearbox with inspection cover removed

2.7 Pull off the end cover complete with ratchet assembly

3   Dismantling the gearbox - removing the gear clusters and camplate

1   Before further dismantling can take place, it is necessary to remove the clutch from the end of the mainshaft and the final drive sprocket from the sleeve gear. Refer to Chapter 1, Section 8 for the appropriate procedure. The engine can be locked during the dismantling operation by selecting top gear and applying the rear brake, preferably before the gearbox inner cover is removed.

2   The final drive sprocket is secured by a large diameter nut with a LEFT HAND thread and a locking plate and grub screw. Remove the grub screw and locking plate, then prevent the sprocket from turning whilst the nut is slackened by locking the back wheel as above. A 1½ inch AF set spanner is needed to fit the nut.

3   Reverting to the other end of the gearbox, remove the inner cover and withdraw the low gear pinion from the end of the mainshaft and unscrew the selector fork spindle. The selector forks can now be disengaged from the camplate and withdrawn, followed by the mainshaft itself complete with gear cluster.

4   Withdraw the layshaft complete with gear cluster, then

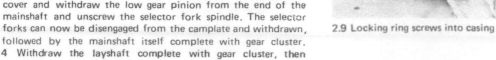

2.9 Locking ring screws into casing

2.11 Gearbox mainshaft nut has right hand thread

2.12 Inner cover pulls off (Note: Mainshaft attached in this photo)

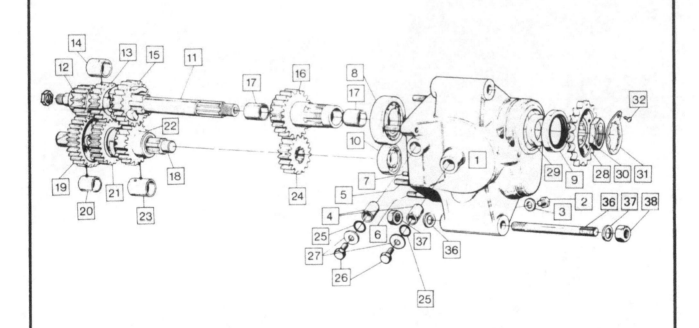

Fig. 2.2. Gear cluster and gearbox shell

1 Gearbox shell with bushes and studs
2 Drain plug
3 Drain plug washer
4 Bush, quadrant and cam spindle 2 off
5 Stud for inner case 2 off (Short)
6 Stud for inner case 5 off (Long)
7 Dowel 2 off
8 Sleeve gear bearing
9 Sleeve gear bearing oil seal
10 Layshaft bearing
11 Mainshaft
12 Mainshaft 1st gear
13 Mainshaft 2nd gear
14 Mainshaft 2nd gear bush
15 Mainshaft 3rd gear
16 Sleeve gear complete with bushes
17 Sleeve gear bush 2 off
18 Layshaft
19 Layshaft 1st gear with bush
20 Layshaft 1st gear bush
21 Layshaft 2nd gear
22 Layshaft 3rd gear
23 Layshaft 3rd gear bush
24 Layshaft 4th gear
25 'O' ring 2 off
26 Spindle bolt 2 off
27 Spindle bolt washer 2 off
28 Gearbox sprockets (17-24 teeth sizes available)
29 Gearbox sprocket spacer
30 Gearbox sprocket nut
31 Sprocket nut lockwasher
32 Gearbox sprocket nut lockscrew

remove the mainshaft sleeve gear. It is a tight fit in the gearbox
main bearing and should be displaced by driving it inwards into
the gearbox shell, through the main bearing, using a rawhide
mallet to obviate the risk of damage.

5   Unscrew and remove the acorn-shaped nut from the front
underside of the gearbox shell. This contains the spring-loaded
camplate plunger which will be exposed when the nut is
removed. Remove the two bolts and washers which secure the
camplate and quadrant to the gearbox shell and lift these
components out. Do not lose the knuckle pin roller in the end of
the quadrant assembly .

3.2 Sprocket nut has a left hand thread

3.3 Unscrew the selector fork spindle

3.4 A side view of the gear cluster

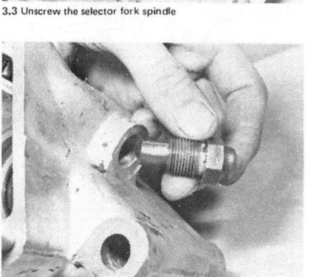

3.5a The camplate plunger

3.5b Complete assembly is secured to gearbox shell by two
bolts on exterior of shell

4 Dismantling the gearbox - removing the gearbox bearings

1   The gearbox has four bearings, two ball journal bearings in the left hand side of the gearbox shell to support the mainshaft sleeve gear and layshaft respectively. The gearbox inner cover contains a further ball journal bearing for the right hand end of the mainshaft; the right hand end of the layshaft is supported in a bush fitted within the kickstarter shaft.

2   The ball journal bearings and the bush are an interference fit within their respective housings. If is necessary to heat the housing in each case so that the bearing can be drifted out of position, or if the housing is blind, by bringing the housing down sharply on a flat wooden surface so that the bearing is displaced by the shock. It is essential that heat is applied before any attempt is made to remove a bearing. If difficulty is encountered in removing this bearing, the gearbox should be removed from the frame.

5 Removing the gearbox as a complete unit

1   As mentioned previously, the gearbox cannot be removed on its own. If the gearbox complete is required out of the frame, the engine and gearbox unit must be removed together as in Chapter 1. Sections 1 - 9.

2   The gearbox is held to the main engine plates with two long bolts. The bottom stud also retains the primary chaincase. Always ensure that the LEFT HAND NUT IS REMOVED so that the chaincase may be taken off ie. the bolt cannot be removed from the frame until the engine/gearbox unit is lifted.

3   The upper bolt has a shouldered head. This shoulder engages with the slot in the left hand engine plate. When tensioning the primary chain, it is the nut on this bolt only which requires slackening before adjusting the drawbolt.

6 Examination and renovation - general

1   Each of the various gearbox components should be examined carefully for signs of wear or damage after they have been cleaned thoroughly with a petrol/paraffin mix. A cleansing compound such as Gunk or Jizer, is particularly useful if the gearbox castings are covered with a film of oil and grease. Make sure all the internal parts have been removed if the gearbox has been dismantled prior to treatment, otherwise the subsequent water wash will cause reuting and damage to the bearings.

2   All gaskets and oil seals should be renewed, regardless of their condition, if the rebuilt gearbox is to remain oiltight. A rag soaked in methylated spirits provides one of the best means of removing old gasket cement without having to resort to scraping and risk of damaging the mating surfaces.

3   Check for any stripped studs or bolt holes which must be reclaimed before reassembly. Internal threads in castings can often be repaired cheaply by the use of what is known as a 'Helicoil' thread insert, without need to tap oversize. Many motor cycle repairers can offer a 'Helicoil' service.

7 Gear pinions, selector arms and bearings - examination and renovation

1   Examine each of the gear pinions to ensure there are no chipped, rounded or broken teeth and that the dogs on the ends of the pinions are not rounded. Worn dogs are a frequent cause of jumping out of gear; renewal of the pinions concerned is the only effective remedy. Check that the inner splines are in good condition and the pinions are not slack on the shafts. Bushed pinions require special attention in this respect, since wear will cause them to rock. Make sure there is no damage to the case hardening of the gearbox pinion teeth.

2   Check both the layshaft and the mainshaft for worn splines, damaged threads and other points at which wear may occur, such as the extremities which pass through the bearings. If signs of binding or local overheating are evident, check both shafts for straightness .

3   Examine the selector forks to ensure they are not twisted or badly worn. Wear at the fork end will immediately be obvious; check the arm in conjunction with the gear pinion groove with which it normally engages. Do not overlook the pin which engages with the camplate track; this is subject to wear.

4   The three ball journal bearings should be washed out with petrol, lightly oiled, and examined for damaged ball tracks. Reject any bearing which has more than just perceptible side play, or is noisy or rough when rotated. Check the bush within the kickstarter spindle with the first gear in position, so that the layshaft is in its normal running position. Renew if the layshaft is a slack fit.

4.2 Never try to remove this bearing (or the one above it) without detaching the gearbox from the machine

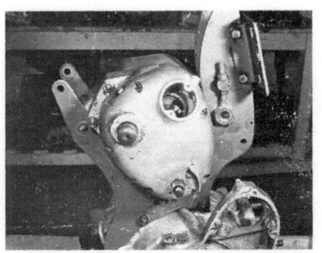

5.2 The gearbox ready for removal from the engine plates

7.4 Drift out the mainshaft bearing after heating casing

8 Kickstarter ratched and gear selector mechanism - examination and renovation

1   The kickstarter ratchet is cut within the outer face of the layshaft bottom gear pinion and is engaged by means of a spring--loaded pawl attached to the kickstarter shaft. After a lengthy period of use, the edges of the ratchet teeth will wear in conjunction with the edge of the pawl and the kickstarter will show a tendency to slip under heavy load. If the ratchet teeth show signs of wear on their leading edges, renewal of the bottom gear pinion will be necessary, also the pawl. It is a wise precaution to renew the pawl spring and pawl pivot pin on the same occasion, especially since it is a low cost item.

2   Check the kickstarter return spring, which is coiled around the kickstarter shaft, has not weakened or stretched. It is a wise precaution to renew this spring too, whilst the gearbox is dismantled.

3   The gear change mechanism rarely gives trouble, apart from the occasional breakage of the gear change lever return and/or ratchet springs. Both springs should be examined and replaced if there is any doubt about their condition.

4   The gearbox camplate should be examined, especially if the pins of the selector forks have worn. The tracks in the camplate are subject to wear after an extended period of service, wear that will be most obvious where the tracks change direction. Wear in the selector mechanism will render gear changes less precise and will cause a general sloppiness of the gear change lever.

5   Do not overlook the camplate plunger and spring. If the plunger jams in its housing or the tension spring weakens, the gearbox will tend to jump out of gear, since the gear which is engaged will no longer be retained positively. Make sure that the acorn nut which forms the housing for the plunger and spring is tightened fully, otherwise some of the spring pressure will be lost.

9 Gearbox shell and end covers - examination and renovation

1   Examine the gearbox shell and the inner and outer covers for cracks or damaged mating surfaces. Small cracks will require expert attention, since they can probably be repaired by welding. Larger or more extensive cracks will necessitate the purchase of a replacement. It is not practicable to effect a satisfactory repair to a badly cracked casting in view of the distortion that may occur.

2   Damaged mating surfaces will cause oil leaks and if the indentations or marks are deep, renewal of the casting will be necessary. Small imperfections can often be sealed off if a liberal coating of gasket cement is used in the area affected, in combination with a new gasket. Beware of oil leakage from the

gearbox since the oil may reach the rear tyre and cause the rear wheel to lose adhesion with the road surface. If in doubt, always play safe and renew the defective part.

10 Gearbox reassembly - refitting the gearbox bearings, gear clusters and selectors

1   Heat the gearbox shell in order to expand the bearing housings and press both the mainshaft sleeve gear bearing and the layshaft bearing into their correct locations. Make sure both bearings enter their housings squarely and are driven fully home.

2   Working from the outside of the gearbox, fit a new oil seal into the bearing housing, lipped side towards the sleeve gear bearing.

3   Locate the gear change quadrant within the gearbox shell bearing and replace the retaining bolt and washer,which should be tightened fully. Do not omit the O ring seal which must be replaced, if damaged.

4   Raise the knuckle of the quadrant until the uppermost end of the curved portion is in line with the top right hand stud of the gearbox shell (see Fig. 2.3). Whilst the quadrant is retained in this position, fit the camplate so that the teeth of the quadrant engage with the teeth at the rear of the camplate. The smooth edge of the camplate should face outwards and the last indentation in the other portion of the camplate should be positioned immediately above the orifice of the camplate plunger. If the quadrant and camplate are lined up in this position, the gear change mechanism is 'timed' correctly and the gears will select in the correct sequence.

5   Replace the bolt and washer which retain the camplate in position, not omitting the O ring seal. Replace the camplate plunger and tighten the acorn nut fully. Check that the alignment of the camplate in relation to the quadrant is still correct.

6   Fit the sleeve gear, complete with bushes, through the sleeve gear bearing, spacer and oil seal, taking care not to damage the latter. Coat the inside of the oil seal and the projecting shaft of the sleeve gear to obviate the risk of damage, whilst the sleeve gear is driven through the bearing from within the gearbox shell.

7   Insert the mainshaft through the sleeve gear pinion and fit the layshaft third gear pinion and bush to the layshaft, followed by the layshaft fourth gear pinion, flat side facing inwards, towards the 3rd gear pinion. The layshaft, complete with pinions, can now be pushed into the layshaft main bearing, within the gearbox shell.

8   Assemble together the mainshaft third gear pinion and the selector fork, then slide the assembly along the mainshaft and engage the pin of the selector fork with the innermost track of the camplate. Replace the mainshaft second gear, complete with centre bush, on the mainshaft. The dogs on the end of this pinion should face the inside of the gearbox.

9   Assembly together the layshaft second gear pinion and selector fork, slide the assembly along the layshaft and engage the pin of the selector fork with the outer track of the camplate. The selector rod can now be inserted through both selector forks and threaded into the gearbox shell. The rod has a flat on the outer end so that it can be tightened fully with a spanner.

10 Fit the layshaft first gear pinion and the mainshaft bottom gear pinion, the latter with its long extension facing outwards.

11 Gearbox reassembly - refitting the inner cover

1   Before refitting the inner cover of the gearbox, make sure the roller is replaced in the end of the gear change quadrant arm. It is not possible to fit the roller after the inner cover has been located since projections in the casting prevent its insertion.

2   If the mainshaft ball journal bearing has been displaced for renewal, the inner cover casting must be heated before the new replacement is fitted. Check that the kickstarter spindle, if removed, is replaced so that the pawl is behind the stop on the inner cover, as shown in Fig. 2.4.

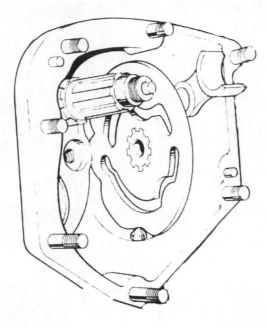

Fig. 2.3. When reassembling gearbox, knuckle end of quadrant must be in line with the top right hand stud of the gearbox shell. This will ensure correct 'timing' of gears.

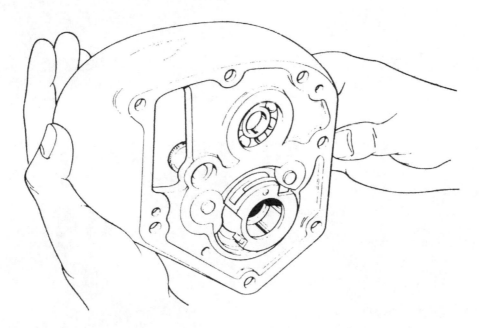

Fig. 2.4. Kickstarter pawl stop must locate with inside of inner cover, as shown

10.2 Don't forget to renew the sleeve gear bearing oil seal

10.4 Replace the quadrant and camplate. Position as in Fig. 2.3.

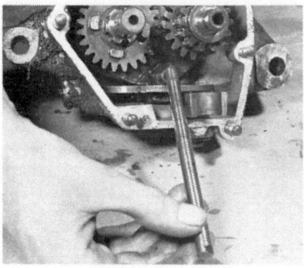
10.9 Screw in the selector rod

10.10 The gear cluster and selectors assembled

11.4 The knuckle roller must be in position, before the inner cover is fitted

11.7 Replace the mainshaft nut

3 Fit a new jointing gasket and ensure the dowel pins are in position in the gearbox shell. Gasket cement is not needed if a genuine Norton gasket is used.

4 Check that all the gearbox components are pushed home fully, especially the selector rod, which must engage to the full depth of thread. Then fit the inner cover, checking that it locates with the dowels and that the quadrant roller is still in position. It may be necessary to guide the unsupported end of the selector rod into position and to twist the inner cover to and fro a small amount so that the dowels will locate correctly. When the cover is fully home, replace the seven nuts which retain the end cover, but before they are tightened, check that both the layshaft and the mainshaft revolve quite freely. If they bind or if the inner cover will not seat correctly, one of the gearbox components has not located correctly, or assembled in the correct order, or the kickstart pawl is not behind its stop. If the end cover is tightened under these conditions it may crack or become permanently distorted.

5 When the gearbox shafts revolve freely, tighten the seven nuts which retain the inner cover to a torque setting of 10-15 ft lb.

6 If the gearbox has been removed from the frame, the engine and gearbox unit should be re-assembled at this point and replaced in the frame. (See Chapter 7, Sections 27 and 29).

7 Replace the mainshaft nut on the right hand end of the gearbox mainshaft. Do not torque up at this point.

8 Fit the spacer (all models with AMC gearbox) for the final drive sprocket and then the final drive sprocket itself. The sprocket has a splined centre fitting which engages with the end of the sleeve gear shaft and is retained by a large diameter nut with a LEFT HAND thread. Reconnect the final drive chain, (closed end of the spring link facing in the direction of motion), and apply the rear brake to lock the mainshaft during the tightening operation. Tighten to a torque setting of 80 lb ft and fit the locking plate and screw.

9 With the gearbox still locked, tighten the main shaft nut (in the right hand casing) to torque setting of 70 ft lb.

10 Replace the kickstarter return spring on the kickstarter shaft. The spring is positioned so that the projection portion faces outwards; the spring should be slid down the kickstarter shaft until the forward facing end enters the locating hole in the shaft. Check that the pawl is located correctly, as detailed previously, then tension the kickstarter spring by turning the spring clockwise until it locates with the stop pin to the right of the kickstarter shaft housing. A length of string to pull the spring round is ideal. Pre-1964 models have a hole in the casting in which the spring will locate.

11 If the clutch withdrawal lever assembly has been dismantled, which is rarely necessary, reassemble the lever, roller, bush and pivot screw and tighten the locknut. Grease and replace the clutch pushrod, then locate the clutch withdrawal lever and locking ring, not forgetting to insert the large diameter ball bearing first. This bears directly on the pushrod end and is actuated by the clutch withdrawal lever.

12 Tighten the locking ring whilst the clutch withdrawal lever assembly is held in its correct location. The centre punch marks made when the assembly was dismantled will ensure correct re-location (reference Fig 2.5). Tighten the locking ring fully, using Loctite.

12 Gearbox reassembly - refitting the outer cover

1 If the gear change mechanism within the outer cover has been dismantled, a check should be made to ensure the spring retaining washer is fitted between the pawl carrier assembly and the outer cover itself. The pawl carrier assembly must be able to move freely and the ratchet spring must be located correctly. The outer cover can be fitted with the ratchet plate assembly attached, or inserted into the already reassembled inner cover; the method of reassembly is identical.

2 Fit a new jointing gasket. Check that the locating dowels are

11.10 Install and tension the kickstart return spring (post-1963 spring shown)

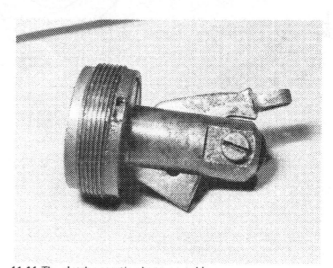

11.11 The clutch operating lever assembly

11.12 Screw in the assembly locking ring

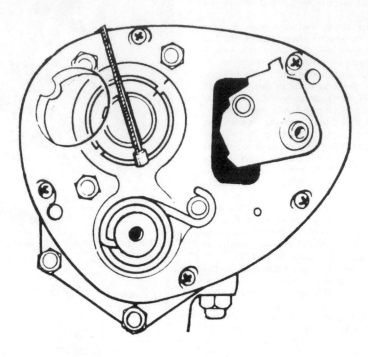

Fig. 2.5 Clutch body must be aligned to give straight pull on cable. Note location of kickstarter spring (post-1964 models)

fitted to the inner cover. Gasket cement is not required when genuine Norton gaskets are used.

3 Locate the ratchet spring in the central position and guide the outer cover over the kickstarter shaft, so that it engages with the dowels. If the cover shows any reluctance to locate fully, it is probable that the pawl has rotated, because the ratchet spring was not located correctly and has permitted the pawl to move. It will be necessary to withdraw the outer cover and reset the pawl and spring before making another attempt at refitting. Alternatively, assemble the pawl carrier in the outer cover, using an outer cover screw in the gear position indicator hole, and align the carrier with the knuckle pin roller hole.

4 When the outer cover is fully home, replace and tighten the five screws which retain the cover in position. Refit the kickstarter and gear change lever and check that both operate correctly. It is important that the pinch bolts of both levers are tightened fully, otherwise the levers will work loose and damage the splines.

5 Check the gear selecting action by engaging each gear in turn whilst rotating the rear wheel to facilitate the engagement of the dogs. If the gears select in a positive and satisfactory manner, check that the gearbox drain plug has been refitted and is tightened fully, then refill the gearbox with oil through the clutch lever inspection cover. It holds 1.0 imperial pints of SAE 90 EP oil. Replace the level plug. Reconnect the clutch cable making sure the nipple engages correctly with the clutch withdrawal lever, then replace the inspection cover, using a new gasket. The cover is secured by two screws.

6 Before taking the machine on the road for the initial test run, verify that the clutch adjustment is correct. It is most important

that there is no loading on the clutch pushrod, otherwise overheating will cause the hardened ends of the rod to soften and adjustment to require frequent attention. For clutch adjustment refer to Chapter 3.7.

---

13 Primary chain - examination, lubrication and adjustment

1 In theory, the primary chain runs under near ideal conditions, in the sense that it is totally enclosed and is running in an oil bath, it will nonetheless require periodic attention to take up any slackness that has developed as the result of wear.

2 Adjustment is effected by drawing the gearbox backwards in the rear engine plates, so the distance between the centres of the engine sprocket and the clutch sprocket is increased. A chromed inspection cap on the primary chaincase provides a convenient means of checking that the chain tension is correct end of examining the chain if it is suspected that any damage has occurred. (The G15CS has a threaded cap in the outer alloy chaincase).

3 Chain tension is correct if there is a movement of 3/8 inch on 750cc models (½ - ¾ inch on smaller engines), in the middle of the top run of the chain, when it is at its tightest point. It is advisable to measure the amount of play when the engine is in several different positions, because a chain rarely wears in an even manner. If adjustment is necessary, slacken the top gearbox locating bolts and use the drawbolt adjuster on the right hand side, immediately above the gearbox, to draw the gearbox backwards until the chain is just on the tight side then use the

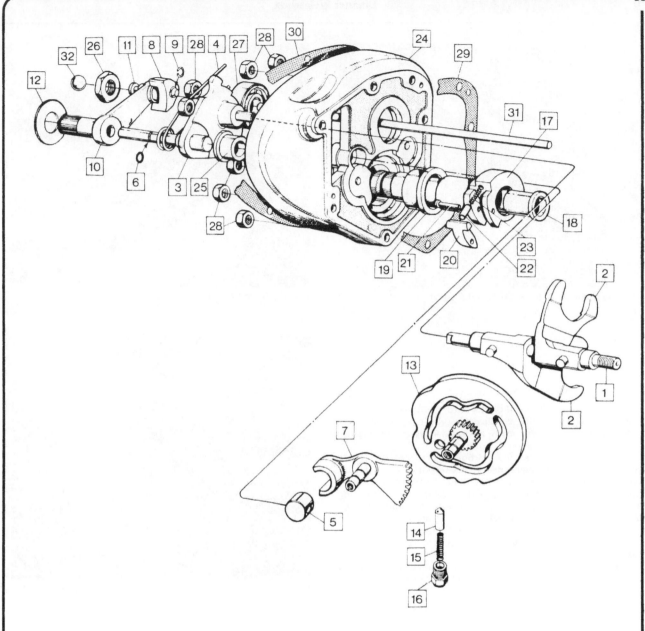

FIG. 2.6. Positive stop and gear selector mechanism

1 Selector fork spindle	11 Pawl pivot pin	*kickstarter*	27 Mainshaft bearing
2 Selector fork 2 off	12 Washer for spring	20 Kickstarter pawl	28 Gearbox inner cover
3 Ratchet plate assembly	13 Camplate	21 Kickstarter pawl pin	stud nut (7 off)
4 Ratchet spring	14 Cam plunger	22 Kickstarter pawl	29 Gasket, inner cover to
5 Knuckle pin roller	15 Plunger spring	plunger	gearbox shell
6 Ratchet spindle 'O' ring	16 Plunger spring bolt	23 Kickstarter pawl	30 Gasket, inner cover to
7 Quadrant	17 Kickstarter shaft with	spring	outer cover
8 Gearchange pawl	bush	24 Gearbox inner cover	31 Clutch pushrod
9 Gearchange pawl circlip	18 Kickstarter shaft bush	25 Gearchange inner bush	32 Clutch operating ball
10 Pawl carrier assembly	19 Inner cover bush for	26 Mainshaft nut	

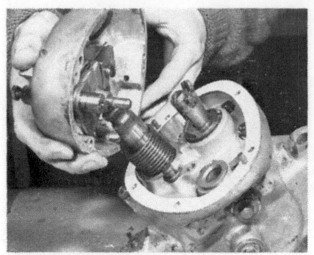

12.3 Use new gasket when fitting outer cover

12.5a Refill the gearbox

drawbolt to move the gearbox forward to obtain the correct
chain tension. Note: this enables the drawing mechanism to
support the gearbox and stops hard acceleration drawing the
gearbox back, thus overtensioning the chain.

4   On the GI5CS model, the gearbox adjusting mechanism is
situated between the engine plates; slacken off the top engine
plate bolt and unlock the adjusting bolt. To tighten the chain,
screw in the adjusting bolt and jerk the gearbox back by pressing
hard on the bottom run of the rear chain. Now screw out the
adjusting bolt until the correct tension is achieved. Lock the
adjusting bolt and tighten the engine plate bolt. Recheck tension
of both chains. In both cases it is advisable to again recheck the
tension after the first run.

5   Retighten the gearbox bolt and lock the drawing mechanism
to the support; recheck the primary chain tension. Note: any
adjustment to the primary chain will alter the tension of the
final drive chain which must be re-tensioned accordingly.

6   To check the primary chain for wear, first wash it free of oil
with a petrol/paraffin mix then compress the chain endwise so
that all play in the rollers is takun up. Make a mark at each end,
then anchor one of the ends and pull the chain in the opposite
direction so that all play is taken up in the opposite direction. If
the chain extends by more than 1/8 inch, it is due for renewal.

7   If the machine is used at irregular intervals, or for a series of
short journeys, the oil in the primary chaincase should be
changed at more frequent intervals than recommended by the
routine maintenance instructions. This will help offset the
effects of condensation, a rusty brown deposit on the chain links
which indicates the oil is becoming contaminated.

8   Use only a good quality chain. Although cheap chains are
available they have a much shorter working life in many cases.
Remember that a broken chain can cause extensive damage,
apart from the possibilty of locking the primary transmission
without warning.

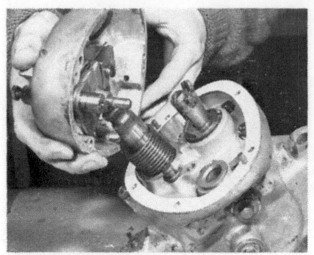

12.5b Replace the level plug when all excess oil has drained off.

---

14 Changing the gearbox final drive sprocket.

---

1   Although the manufacturer has selected gear ratios that give
optimum performance and it is therefore difficult to improve
overall performance as the result, occasions will arise when a
change has to be made, such as when engine performance is

increased by tuning. Proprietary manufacturers can supply a
range of gearbox final drive sprockets for the standard road
models, available for 5/8 x 1/4 in chain up to 19 teeth.

2   From the preceding Sections of this Chapter, it will be
appreciated that a certain amount of dismantling is necessary in
order to effect a sprocket change, namely the removal of the
primary transmission and chaincase, as described in Chapter 1.8.
In consequence, a change in sprocket size is not a matter to be
taken lightly, especially since the whole process will need to be
repeated if the end result is not satisfactory. It is therefore
advisable to seek the advice of a Norton Villiers specialist before
contemplating such a change; the fitting of a larger final drive
sprocket by no means guarantees an increase in maximum speed.

3   The gearbox final drive sprocket should be examined at
intervals to check for wear, chipped or broken teeth or other
defects which will otherwise cause rapid chain wear and harsh
transmission. A badly worn sprocket should be renewed at the
earliest possible opportunity as the rate of wear is 2-3 times that
of the rear wheel sprocket due to epicyclic variation.

15 Fault diagnosis

Symptom	Cause	Remedy
Kickstarter does not return when engine is turned over or started	Broken kickstarter return spring	Replace.
Kickstarter slips and will not turn engine over	Worn ratchet assembly	Replace.
Kickstarter jams	Worn ratchet assembly and/or pawl	Replace.
Difficulty in engaging gears	Gear selectors not indexed correctly	Check alignment of 'timing' marks on inner cover.
	Selector forks bent or badly worn	Replace.
Machine jumps out of gear	Camplate plunger sticking Worn dogs on gear pinions	Remove and free. Replace defective pinions.
Gear change lever does not return to original position	Broken return spring	Replace spring in outer cover.
Harsh transmission and high vibration levels with feeling of power loss.	Primary and/or final drive chain too tight	Check tension of chains; check especially for tight spots
Kickstart suddenly drops down when accelerating	Layshaft ball bearing failed	Renew bearing, and bush inside layshaft

Chapter 3 Clutch

Contents

Specifications

Clutch

Type	Multiplate
Number of Friction Plates	5 + End Plate
Number of Plain Plates	5

Note: "See Fig 1.5 for breakdown of clutch"

1 General Description

The clutch fitted to all the Norton twins covered in this manual is virtually indentical and has been universal on all large capacity Nortons since the 1930's. The only major development of this unit was the inclusion of bonded friction plates in the late 1950s; these plates replaced the cork insert type of clutch plate and necessitated certain design changes. The original design of clutch sprocket contained cork inserts and, when the bonded inserts were introduced it became necessary to reverse the relative positions of the inserted and plain plates.

This multi-plate clutch with a hardened steel centre and three separate clutch springs is of conventional motorcycle design; its "larger than normal" diameter has made it very popular for 'special' builders, along with the AMC/Norton gearbox.

2 Dismantling the clutch

1 Access to the clutch is gained by removing the primary chaincase. This will require the removal of the left hand footrest, the slackening of the rear brake and the removal of the centre sleeve nut (see Chapter 7 Section 8). Since the chaincase does not have a drain plug, a container should be placed under the chaincase before the sleeve nut is slackened. When the inner and outer chaincases are parted, approximately 130cc. of oil will be released.

2 If only the clutch plates require attention, there is no need to dismantle the primary transmission. The clutch plates are readily accessible with the clutch chainwheel in situ.

3 The 3 clutch springs are retained by shouldered nuts. These should be removed with a divided screwdriver although a small

screwdriver on one side only will probably suffice if the nut head is not worn.

4 As with all mechanical parts which have relative movement, it is advisable to dismantle the clutch so that the mating parts may be replaced in their previous positions. Withdraw the clutch plates from the clutch sprocket as a unit after removing the clutch springs, the spring cups and the outer, alloy, pressure plate. Alternatively, if the clutch sprocket and centre are to be removed, it is necessary to slacken off the gearbox and remove the primary chain spring link and remove the chain from the clutch. In this case, the clutch plates and clutch sprocket may be withdrawn complete from the centre.

3 Clutch plates - examination and renovation

1 Check each clutch plate to ensure it is completely flat and free from any blemishes. Reject any that are distorted, or clutch troubles will persist. Note that the plain clutch plates which mate with the bonded surfaces are of the 'pin point planished type' ie. they have small pop marks all over.

2 Examine the teeth at the edge of the plain clutch plates and in the centre of the friction plates. It is important that they should be in good condition and free from any burrs or other

2.5 Remove pressure plate and springs to withdraw clutch plates complete

damage. Burrs can be removed by dressing with a file; if any teeth are chipped or broken, the clutch plate should be replaced.

3 Clutch slip at high speeds is usually an indication that the thickness of the friction plates is at the lower limit. Always replace the friction plates as a complete set, never singly, except on the early clutches which use cork inserts. In this case renew a maximum of 3 plates at one time or it may be difficult to disengage the clutch until the new plates have bedded in.

4 Clutch slip will also occur if the friction plates are soaked in oil. Wash in highly volatile solvent (not petrol or parrafin) and dust with Fullers Earth to absorb the oil.

4 Clutch inner and outer drums - examination and renovation

1 Examine the slots with which the clutch plate teeth engage. If they are indented badly, renewal of the inner and/or outer drum will be necessary. Minor indentations can be dressed with a file so that the slots have parallel sides again. Clutch drag and general uncertainty of clutch action can usually be attributed to such indentations which trap the plates and restrict their

freedom of movement.

2 Indifferent clutch action can be attributed to a worn clutch roller cage. Wear in this component may be checked when the clutch unit is fully assembled (without the primary chain). Grip the clutch body and attempt to rock it sideways. If excess movement is evident, the clutch rollers and/or the race plate will require replacing. Refer to Chapter 1.8, paragraph 12 for advice about the removal of the clutch centre. Note: The clutch shock absorber centre fits in the clutch backplate and will require separating before the clutch centre body may be removed. Note that if the clutch centre nut has slackened, this can give the impression that the bearing has failed.

3 The clutch centre contains three large and 3 small shock absorber rubbers. Jerky take up on the transmission may be attributable to the decomposition of these rubbers from the ingress of oil. Access to the rubbers is achieved by removing the three countersunk screws retaining the steel cover plate and prising out the cover plate; it is a tight fit and a screwdriver 'twist' through each of the spring stud holes in turn will lift the plate.

4 Unless the rubbers are in a very bad state, they will be difficult to remove. It is advised that in this case the replacement of the rubbers is left to a Norton dealer with the correct service tool. When fitting new rubbers, smear them with liquid detergent first, so that they will slide into position more readily.

5 The clutch chainwheel should be examined, checking for general wear, chipped or broken teeth. If damage of this nature is found, the complete clutch outer drum must be renewed since the chainwheel is an integral part.

5 Clutch spring examination

1 The clutch springs should be checked against each other for length. If any discrepancy occurs, all three should be renewed.

6 Reassembling the clutch

1 Reassemble the clutch by reversing the dismantling procedure detailed in Section 2 of this chapter. Make sure that the clutch plates are alternated correctly. Do not forget to replace the clutch pushrod, which may have been withdrawn during any dismantling operation.

Also replace the clutch adjuster and locknut, after positioning a dab of grease on the end of the adjuster bolt where it makes contact with the pushrod. Adjust the clutch according to the procedure recommended in the following section.

4.3 Remove the three countersunk screws and prise out the cover plate

4.4 The shock absorber showing the rubber inserts

6.2 Screw in the nuts so that they are in line with the stud tops

2   On reassembling the clutch, the spring retaining bolts should be screwed in so that the top of the nut is in line with the stud. Disengage the clutch using the handlebar lever and turn the clutch centre with the kickstart. If the clutch does not spin evenly, adjust each spring nut $\pm$ ¼ turn to eliminate any drag.

3   Engage the clutch, connect up the primary chain and kick over the engine. If the clutch slips, either new bonded plates or springs are required (assuming the clutch cable and pushrod are operating correctly).

7 Clutch adjustment - general

1   In order to adjust the clutch correctly, the following procedure must be followed. Commence by checking there is no tension on the clutch cable. Then screw in the adjuster in the clutch centre until the clutch operating lever loses all its free play. Slacken the adjuster bolt back ½ to ¾ turn and hold it in this position whilst the locknut is tightened. Then re-adjust at the point where the cable enters the gearbox until there is about 1/16 inch to 3/32 inch free play before the cable is under tension.

2   It is most important that the amount of free play in the clutch cable is maintained. If the clutch pushrod is permanently in contact with the clutch actuating mechanism, the continuous loading will cause the rod to heat up and the hardened ends soften. Rapid wear of the pushrod will follow, necessitating frequent clutch adjustment.

3   When making the initial adjustment, occasions occur when the clutch withdrawal lever is displaced, rendering the clutch inoperative. Under these circumstances it is necessary to remove the inspection cover from the right hand side of the gearbox, slacken back the adjuster in the clutch centre and lift the lever back into its correct location before adjusting the clutch again. When replacing the inspection cover, do not omit the sealing gasket.

4   On later models, a hole in the primary chaincase facilitates the adjustment of the clutch without the necessity of removing the primary chaincase. This hole is filled by a rubber bung on 'Dominators' and a threaded alloy cap on the G15CS. For early Dominators, it is recommended that the rubber bung is purchased and a hole drilled to suit in the dome of the outer chaincase. This will permit access to the adjuster, without need to drain the oil content, and remove the outer cover.

5   Replace all parts of the primary transmission, re-adjust the primary chain tension, replace the outer chain cover and refill it with approximately 130 cc s of oil.

8 Fault diagnosis

Symptom	Cause	Remedy
Engine speed increases but not road speed	Clutch slip; incorrect adjustment worn linings or worn clutch springs	Adjust or replace clutch plates/clutch springs
Machine creeps forward when in gear;	Clutch drag; incorrect adjustment or damaged clutch plates	Re-adjust or fit new clutch plates
Machine jerks on take-off or when changing gear	Clutch centre loose on gearbox mainshaft Worn clutch centre rubbers	Check for wear and retighten retaining nut Replace shock absorbing rubbers
Clutch noisy when withdrawn	Badly worn clutch roller race	Renew race and/or rollers
Clutch neither frees nor engages smoothly	Burrs on edges of clutch plates and slots in clutch drums	Dress damaged parts with file if damage not too great
Clutch action heavy	Dry operating cable or bends too tight	Lubricate cable and re-route as necessary
Clutch action harsh	Overtight primary chain	Re-adjust primary chain
Clutch 'bites' at extreme end of lever movement	Worn linings	Replace clutch friction plates
Constant loss of clutch adjustment	Worn pushrod due to failure to maintain minimum clearance	Replace pushrod and re-adjust
	Mainshaft nut loose at either end	Tighten, using Loctite thread sealant.

Chapter 4 Fuel system and lubrication

Contents

Specifications

Carburettors

Make	Amal			
Type Pre 1968	Monobloc			
Right Hand	376/289		389/88	
Left Hand	376/288		389/87	
Post 1968	Concentric			
Right Hand	R930		R930	
Left Hand			L930	
Main Jet - Twin carbs	250	270	350	420
Single carb	240 250	320	350	420
Throttle Slide Monobloc	376/3½		389/3	
Concentric	Size 3½		Size 3	
Pilot Jet	25		20	
Needle	0.1065			
Needle position	3			

Note: The above information should be used only as a guide. Carburettor specifications have changed often during the development of the Dominator engine.

Fuel Tank

Capacity	3.62 Imp. galls	4.35 U.S. galls	16.5 litres
Featherbed Type	3.5/8 Imp. galls	4.35 U.S. galls	16.5 litres
Scrambles type	2.5 Imp galls		

Oil Tank

4.5 Imp pts	5.4 U.S. pints	2.55 litres

1 General description

The fuel system comprises a petrol tank which either rests on the top frame tubes (Featherbed frames) or sits astride the top frame tube ("Scrambles" frame). Petrol is gravity fed to the carburettor(s) which may be of the Amal 'Monobloc' (Pre - 1968) or 'Concentric' (Post 1968) type. The latter unit was developed in the early 60s to eliminate the fuel surge problems with the 'Monobloc' carburettor when cornering at speed on long curves and also to replace the 'handing' of the float chamber on twin carburettor machines). Although lever type petrol taps are used on the later machines, earlier models may have "push ON - push OFF" taps; all taps have a built-in gauze filter which projects up into the tank. If two petrol taps are fitted to the tank, only the left hand one should be used; the other will then provide a small quantity of petrol in reserve.

On twin carburettor engines, the throttle slides and the air slides are linked by means of separate junction boxes so that only one cable emerges from the twist grip and the air lever. The air slides are only used on starting when they act as a choke. A large capacity air cleaner may be fitted in front of the oil tank and the battery case; it was an optional extra for the pre-1969 machines and is available for single or twin carburettors.

Lubrication is effected on the dry sump principle, in which a gear-type mechanical pump feeds oil under pressure to the various engine components, via filters and a pressure release valve. Excess oil is pumped from the crankcase back to the oil tank, by a scavenge pump whose capacity is double that of the pressure pump thus ensuring a dry sump.

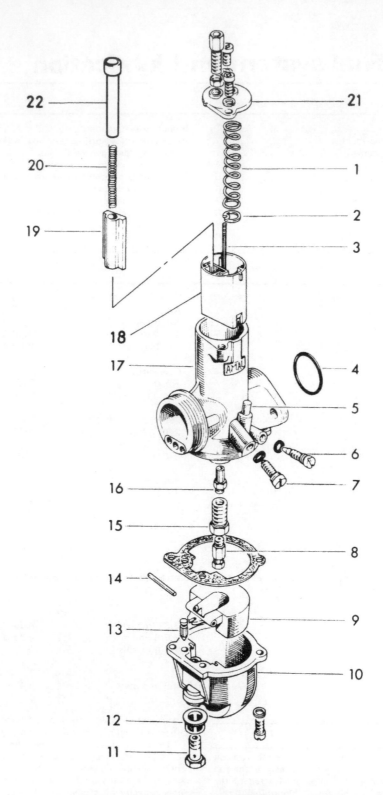

FIG 4.1. COMPONENT PARTS OF THE AMAL CONCENTRIC CARBURETTOR

1 Throttle return spring	7 Throttle stop screw	13 Float needle	19 Air slide (choke)
2 Needle clip	8 Main jet	14 Float hinge	20 Air slide return spring
3 Needle	9 Float	15 Jet holder	21 Mixing chamber top
4 'O' ring	10 Float chamber	16 Needle jet	22 Air slide guide
5 Tickler	11 Banjo union bolt	17 Mixing chamber body	
6 Pilot air screw	12 Filter	18 Throttle valve (slide)	

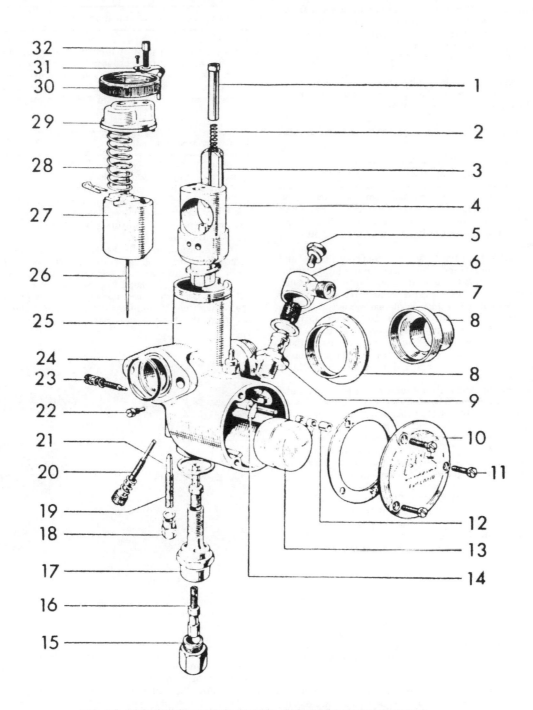

FIG. 4.2. COMPONENT PARTS OF THE MONOBLOC CARBURETTOR

1 Air valve guide	9 Needle setting	18 Pilot jet cover nut	27 Throttle slide
2 Air valve spring	10 Float chamber cover	19 Pilot jet	28 Throttle spring
3 Air valve	11 Cover screw	20 Throttle stop screw	29 Top
4 Jet block	12 Float spindle bush	21 Needle jet	30 Cap
5 Banjo bolt	13 Float	22 Locating peg	31 Click spring
6 Banjo	14 Float needle	23 Air screw	32 Adjuster
7 Filter gauze	15 Main jet cover	24 'O' ring seal	
8 Air filter connection (top or air intake tube	16 Main jet	25 Mixing chamber	
	17 Main jet holder	26 Jet needle	

2   After a long period of service, the transparent plastic material of which the pipes are made will harden and discolour due to the gradual removal of the plasticiser by the petrol. If the pipes are exceptionally rigid, they should be renewed because it is under this condition that they are most likely to crack, especially in cold weather.

3   Never use ordinary rubber tubing, even as a temporary replacement. Petrol causes rubber to swell and disintegrate, thereby blocking the fuel supply completely.

2   Petrol tank - removal and replacement

1   Before the petrol tank is removed from the machine, all petrol taps should be closed and the petrol pipes detached by unscrewing the union joints, at the carburettor. There is no necessity to drain the tank unless it is desired to remove either of the petrol taps.

2   Methods of tank mounting vary, according to the model and the type of petrol tank fitted. "Scrambles" tanks locate with a short metal plate welded across the lower top frame tube, immediately to the rear or the steering head. It is necessary only to remove the locknuts and rubber insulating washers to free the nose of the tank. On the 'slimline' Featherbed frames, the front of the tank is held by two bolts to plates which are welded inside the top frame tube; a rubber pad sits either side of these plates. 'Wideline' featherbed frame tanks are retained by a central strap with a long bolt at the rear which screws into a bracket on the tool tray.

3   At the rear of the petrol tank a rubber ring passes around a metal bracket on the frame and over a hook on the tank. On the scrambles framed models a rubber ring which passes under the frame tube is located with two 'hooks' at the rear of the tank. When either of these fixing methods is released, the tank can be lifted away from the frame. Note the location of the rubber pads on top of the main frame tube which insulate the tank from vibration. They must be replaced in the same position.

4   Replacement is accomplished by reversing the procedure detailed in the preceding paragraphs. Check that the rubber ring is located fully with the 'hooks' and that the vent hole in the filler cap is not obstructed. If the tank is airtight, the supply of petrol will be cut off, leading to a mysterious engine fade-out which is difficult to eliminate without realising the cause.

3   Petrol taps - removal and replacement

1   The petrol tap(s) thread into an insert in the bottom of the petrol tank, one on each side. The taps seat on a fibre washer which should be renewed to obviate leakage, each time the taps are removed and replaced. Most models have only one tap, on the left-hand side.

2   If the rate of flow from the tap(s) becomes restricted, it is probable that the gauze filter within the petrol tank has become choked. Under these circumstances it will be necessary to drain the petrol tank and unscrew the defective tap so that the filter can be cleaned.

3   If the petrol taps leak, they should be dismantled and inspected. Lever taps: remove the nut, washer and spring which retain the tapered barrel; remove the barrel and inspect if deeply scored it must be replaced or else regrind the barrel into the top with a fine abrasive polish. Push ON - push OFF taps; unscrew the two ends of the tap and replace the corks, one either side of the moveable diaphragm. They can be reclaimed by dropping the tap centres, complete with cork, into a cup of boiling water to expand them. Pull ON-push OFF; remove the small retaining screw and pull out the barrel; the cork sealing ring can also be reclaimed by immersion in boiling water.

4   Petrol feed pipes - examination

1   The petrol feed pipes are fitted with unions to make a quickly detachable joint at both the carburettor float chamber and the two petrol taps. Leakage is unlikely to occur unless the union nuts slacken, the tubing splits, the metal ferrules around the pipe ends work loose or the union deforms under overtightening of the retaining bolt (Concentric carburettor).

5   Carburettor - removal

Note: In the following text, (M) denotes Monobloc carburettor (C) denotes Concentric carburettor.

1   Commence by removing the petrol feed pipes at the banjo union joint with the underside of each float chamber (C) or with the top of each float chamber (M). There is a nylon filter within each banjo union which will be displaced when the petrol feed pipe is withdrawn.

2   Where fitted, detach the short rubber hose from each carburettor intake so that the air cleaner is disconnected. Twin carburettor air cleaners are retained by two bolts. Remove the two crosshead screws in the top of each carburettor (C) or unscrew the large knurled ring which retains the top of the carburettor (M) and lift the top away complete with control cables and the throttle valve and air slide assemblies.

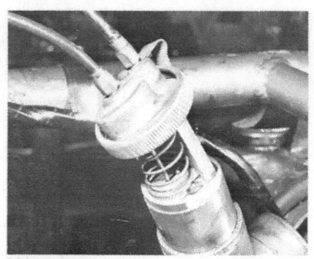

5.2 A large knurled ring retains the top of the Monobloc carburettor

3   Tape the carburettor tops and slide assemblies to some nearby frame member to obviate the risk of damage when further dismantling occurs.

4   Remove the carburettor(s) by unscrewing the nuts which hold the unit to the manifold. Alternatively, on non-'SS' heads with a splayed twin carburettor set-up, it is more convenient to remove the manifold from the head complete with carburettors. Where fitted, do not loose the heat insulating washers or the gaskets. Note: many twin carburettor systems incorporate a small balance pipe between inlet manifolds; ensure that the pipe is in good condition and that the pressure tappings are sealed or the mixture will be 'leaned off' by air leaks.

5   On factory fitted twin carburettor systems, the units are identical in specification apart from the fact that they are 'handed'. This is necessary to ensure the pilot jet screw and the throttle stop screw are always outward facing to facilitate ease of adjustment.

Note: Monobloc systems have their float chambers handed or have the float chamber removed on the right hand unit.

6 Carburettors - dismantling

1 On twin carburettor systems both carburettors are virtually identical and in consequence the same dismantling procedure and layout of the component parts applies to each.

CONCENTRIC

2 Commence by removing the float chamber. This is retained to the underside of the mixing chamber by two crosshead screws and can be lifted away when the screws are withdrawn. It will contain the horseshoe-shaped float, float needle and float spindle. Take care not to damage the gasket between the float chamber body and the base of the mixing chamber.

3 Drain the float chamber, then lift out the float, complete with float needle and spindle. Withdraw both the float needle and the spindle from the float assembly; both these components are small and easily lost. Place them in a safe place for subsequent examination.

MONOBLOC

2(a) Commence by removing the main jet cover and allow the fuel to drain from the float chamber. Remove the float chamber side cover by undoing the three cheesehead screws.

3(a) Remove the spacer on the float support spindle and pull out the float (which may be either brass or nylon). On removal of the float, a nylon or brass needle will drop from the inlet union - do not lose.

4 The main jet can be seen projecting from the base of the mixing chamber. It should be unscrewed, followed by the jet holder above it which has the needle jet threaded into the furthermost end. Unscrew and detach the needle jet. Unscrew the pilot jet.

5 If the throttle stop screw and pilot jet screw are removed from the carburettor, note should be made of their settings so that they can be replaced in approximately the same position. Count the number of turns and part turns from the fully closed position.

6 Reverting to the carburettor tops which have been taped to a frame tube, lift the throttle valve return spring and disengage the spring clip from the needle, after making note of the notch with which the clip was engaged. Then disengage the end of the throttle cable from the throttle valve and lift the valve away.

7 The air slide assembly can be disconnected by disengaging the air cable nipple in similar fashion. However, it is unlikely that this assembly need be disturbed since the air slide rarely requires attention.

8 When dismantling twin carburettors, keep the component parts separate. Problems may occur if parts are unwittingly interchanged, especially moving parts subject to wear.

7 Carburettors - examination of the component parts

1 If the carburettor has shown a tendency to flood, check the float for leaks. Any leakage will immediately be obvious, due to the presence of petrol within the float. It is not practicable to repair a leaking or damaged float; renewal is essential.

2 The float needle and float needle seating should be examined with a magnifying glass. If wear had taken place, it will be evident in the form of a ridge on both the needle point and in the needle seating. This will prevent the needle from shutting off the fuel supply completely and will cause a permanently rich mixture, at the expense of fuel consumption. Renew both needle and seating if this form of wear is evident (C) or only the needle (M). Also renew if the needle is bent or mis-shapen. The float needle seating threads into the float chamber body.

3 Check that the float needle has a Viton rubber tip. This late modification has been found advantageous in overcoming carburettor flooding problems, especially those initiated by high frequency vibration.

4 The float hinge is unlikely to give trouble unless it is badly worn or has been bent as the result of careless assembly. Do not attempt to straighten a bent hinge. It is a low cost item, which should be renewed if damaged (C only).

5 Carburettor jets may block occasionally due to foreign matter which may have been present in the petrol. Never clear a blocked jet with a piece of wire or any sharp instrument since this will enlarge the jet orifice and cause changes in the carburation. Use either compressed air, or a blast or air from a tyre pump.

6 If petrol consumption has shown a tendency to rise, renew both the needle and the needle jet. Wear will occur after a lengthy period of service. Make sure the needle is not bent and that the clip is a good fit.

7 Wear of the throttle valve will be self-evident and may be accompanied by a clicking noise when the engine is running due to the valve rattling within the mixing chamber body. Wear marks are usually found at the base of the valve, on the side nearest the inlet port.

8 Check that neither the throttle stop screw nor the pilot jet screw is bent and that the taper of the latter screw is not worn. Check the threads for soundness and renew the small O rings which form a small but effective seal (C only).

9 If the throttle valve has worn badly, it is probable that the mixing chamber has worn too. Evidence of such wear will be found in the vicinity of the throttle valve bore, close to the inlet and outlet passages.

10 Blow out all the internal air passages of the mixing chamber and check that the internal threads are sound, especially those that accept the crosshead screws retaining the carburettor top and the float chamber (C) or that which mates with the large knurled ring. (M).

8 Reassembling the carburettors

1 Reassemble the carburettors by reversing the dismantling procedure. If the float chamber gasket is damaged, fit a new replacement. Check that it is fitted the correct way round, so that the holes align with the jet passages and the ends of the float spindle are not trapped (C only).

2 When refitting the carburettor top, the needle must engage with the needle jet (the central hole in the jet block), and the air slide with the cutaway in the top of the throttle valve. Do not use force, it is quite unnecessary if everything locates correctly.

3 Be sure to fully tighten the crosshead screws which retain both the carburettor top and the float chamber (C) or the knurled ring which retains the top (M). Do not overlook the cheesehead screws which retain the float chamber side. (M). If the carburettor top works loose, the throttle may jam open, whilst if the float chamber is not secured rigidly, petrol spillage is inevitable.

4 Before reconnecting the petrol feed pipes, check that the nylon filter within the banjo union affixed to the float chamber is clean and not crushed.

9 Carburettors - checking the settings

1 The sizes of the jets, throttle valves, needles and needle jets are predetermined by the manufacturer and should not require modification unless the engine has been tuned.

Use the specifications list at the beginning of this Chapter as a guide. If there is any doubt about the values fitted, check with either Norton or directly with Amal Limited. The author has always found the latter most helpful on all carburation problems.

2 Slow running is controlled by a combination of throttle stop and pilot jet screw adjustment.

Single Carburettor

With the engine running and at normal operating temperature, commence by screwing the throttle stop screw inwards a quarter turn at a time until the machine runs at a fast tick over speed. Adjust the pilot air screw slowly either in or out until the fastest running speed is achieved. Unscrew the throttle stop screw until the desired tick-over speed is obtained, and again check the pilot air screw so that the fastest running speed is achieved. The normal setting for the pilot jet screw is in the region of 1½ complete turns out from the fully closed position. The mixture is enriched by turning this screw INWARDS because it meters the supply of air and not fuel.

These settings should be made only after the machine has been ridden for at least five minutes to warm up the engine evenly.

Twin Carburettors

With the left hand sparking plug removed, set the right hand carburettor as above. Replace the left hand plug and remove the right hand plug, set the left hand carburettor as above. On replacing both plugs and starting the engine, the slow running will probably be too fast; the speed can be lowered by slackening each throttle stop screw an identical amount.

3 As a rough guide, up to 1/8 throttle is controlled by the pilot jet setting, from 1/8 to 1/4 throttle by the throttle valve cutaway, from 1/4 to 3/4 throttle by the needle position and from 3/4 to full throttle by the size of the main jet. These are only approximate divisions; there is a certain amount of overlap.

10 Balancing twin carburettors

1 On machines fitted with twin carburettors, maximum performance can be achieved only if both carburettors are completely in phase with one another. They must commence to open at the same time and must remain in complete synchronisation throughout the entire throttle opening range. Check in the top position first of all in case one throttle stop screw is further in than the other.
2 All checks should be made with a dead engine. Open the throttle slides by turning the twist grip and make sure that both throttle slides commence to rise at the same time. If they do not, adjust the individual throttle cable adjusters until the slides are completely in phase.
3 Check at the other end of the scale by opening the throttle slides fully by turning the twist grip. Neither should obstruct the carburettor intake and as they are lowered they should be completely in step with one another.

11 Induction system joints

1 Whenever the carburettor flange to spacer/manifold joint is broken it is essential to check that the O ring in the centre of each carburettor flange is in good condition and is seating correctly in its groove. An air leak can cause a weak mixture, which may eventually result in a burnt piston or valves.
2 Do not overtighten the carburettor at the flange to spacer/manifold joint. Overtightening will cause the carburettor flange to bow and initiate an air leak which the O ring cannot seal.
3 On single and twin carburettor machines, do not forget the heat insulating inlet tract spacers. They break up the heat path and prevent the carburettors from overheating by conduction.

12 Air cleaner - dismantling, servicing and reassembling

1 The air cleaner chamber contains a corrugated paper element around its periphery and is situated in front of the oil tank and the battery case on single carburettor models.
2 Unless heavily contaminated, wet or soaked in oil, the element can be re-used after it has been tapped on the bench to dislodge any foreign matter and then blown clear with an air line. Although composed of felt, it should be handled with care. If it is torn or perforated, renewal will be necessary.

3 Replace the element by reversing the dismantling procedure. Tighten the retaining bolts after a check has been made to ensure that the element is not trapped at any point.
4 It follows that the carburettor intake hoses should be in good condition, otherwise air leaks will occur. If the machine was supplied with an air cleaner, on no account run the machine with the air cleaner disconnected or without an element. The carburettors are jetted to take the presence of the air cleaner into account and if this advice is overlooked, a permanently weak mixture will result.

13 Exhaust system - general

1 Although many different types of exhaust system may be fitted (see Chapter 1 Section 6), the following general points will apply.
2 Absolute tightness of the finned exhaust pipe nuts is essential. If they work loose, they will chatter away the internal threads within the exhaust ports, necessitating repair by a

13.2 Ensure the finned exhaust pipes are always tight or threads will strip

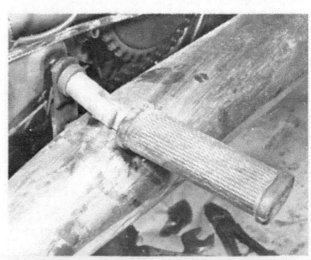

13.5 Silencers locate at pillion footrests on standard road models

Norton specialist. Locktabs should be used to prevent this happening. Damaged exhaust port threads can be reclaimed by threaded inserts or Helicoils.

3 A sealing washer must be located between the exhaust port opening and the end of the exhaust pipe. This will obviate the occurrence of air leaks, which often cause a mysterious and difficult to eliminate backfire which occurs only on the over-run. Genuine Norton sealing washers are made of steel.

4 Various designs of silencer are fitted, some of which conform to the noise reduction requirements of certain overseas countries. Unlike the exhaust system of a two-stroke, the silencers are unlikely to require cleaning out at regular intervals.

5 The silencer mountings (ie. the rear footrest mountings) should be checked at frequent intervals to ensure they are tight.

14 Lubrication system- general

1 The lubrication system functions on the dry sump principle. Oil from a separate side-mounted oil tank is fed by gravity and suction to the feed section of a gear-type pump housed in the right hand crankcase and driven from the right hand end of the crankshaft. The pump delivers oil under pressure through the crankshaft to both big ends via a pressure release valve. Oil escaping from the big end bearings lubricates the cylinder walls, main bearings and camshaft by splash. The inlet rocker box drains via an oilway in the cylinder barrel and the exhaust rocker box through an oilway into the pushrod tunnel, where the return flow of oil lubricates the cam followers. Oil which collects in the crankcase sump is picked up by the scavenge section of the oil pump and returned to the oil tank for recirculation. A small bleed off the main oil supply is used to lubricate the rocker gear on very late models only.

15 Dismantling, renovating and reassembling the oil pump

1 Unless the oil pump gives trouble, it should not be dismantled unnecessarily. Oil pump faults can be divided into two categories; complete failure of the pump and a tendency for oil to leak through the pump whilst the machine is stationary, causing the crankcase to fill with oil. The first type of fault is usually associated with some form of breakage in the oil pump drive, or the presence of metallic particles in the pump itself which has caused the gears to lock, shearing one of the drive spindles. The less serious oil leakage is caused by general wear and tear which can be rectified to an extent by taking up end float, as described in the following paragraph. Any fault attributed to the oil pump necessitates a complete strip down for examination.

2 Remove the oil pump as a complete unit from within the timing chest, as described in Chapter 1.11, paragraph 1. If the feed gear is not a tight fit on the pump main shaft, it will only show wear on the return side of the pump.

3 To strip the pump, remove the four long screws which hold the oil pump together. The top cover can be lifted off together with the drive gear and spindle; the other pump gears can be lifted out, also the second spindle. Only the drive gear (feed side) is keyed on to the main shaft. One is part of the main shaft and the others a sliding fit.

4 Wash all the components in neat petrol and allow them to dry. It will be noted that the scavenge section of the pump is twice the capacity of the feed section, necessary to keep the crankcase clear of excess oil. In consequence, the scavenge pump gears have wider teeth.

5 End float in the pump requires expert attention to be taken up as surface grinding is essential.

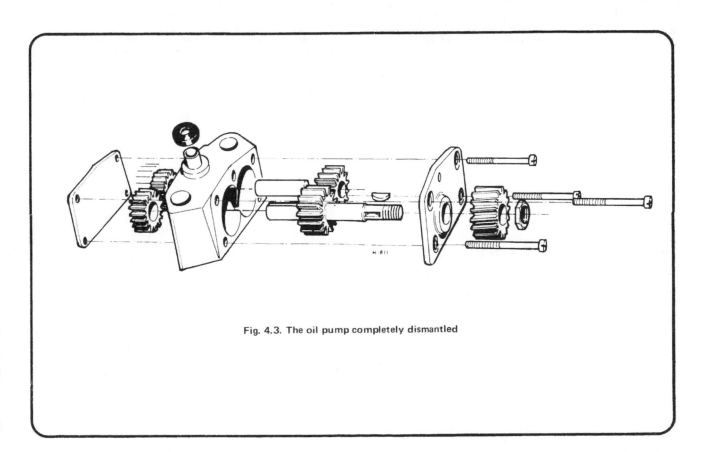

Fig. 4.3. The oil pump completely dismantled

6   Prime the pump with clean engine oil by pumping oil from a pressure oil can into the main feed orifice, whilst turning the driving spindle so that oil will circulate through the gears. The pump should now revolve more freely, the slight stiffness decreasing or even disappearing completely.

7   It follows that if metallic particles were found when the pump was stripped and if any of the gears have chipped or broken teeth, the pump should be renewed as a complete unit. Remember, that if the oil pump does not function correctly, serious engine damage will result.

8   The oil pump driven gear is keyed onto the pump spindle and is retained by a self-locking nut. Replacement of the gear pinion is easy, especially since it is a parallel fit on the shaft.

16 Oil pressure release valve

1   An oil pressure release valve is fitted in the left hand end of the timing cover to prevent the oil pressure from rising above 40 to 50 psi. It comprises a spring-loaded plunger pre-set during manufacture of the machine by inserting shims so that it will actuate at the desired pressure. It requires no attention, other than the occasional check that the dome nut is tight and there is no oil leakage.

2   The normal oil pressure is 30 to 40 psi at 3,000 rpm when starting from cold.

17 Crankcase breather

1   All models have a timed and ported crankcase breather which takes the form of a spring-loaded rotary disc with cutaway segments, driven from the left hand end of the camshaft. The cutaways align with similar cutaways in a stationary disc behind the left hand camshaft bush at a predetermined time, which cannot be changed if the driving dogs or the rotary disc align correctly with the corresponding cutaways in the end of the camshaft.

2   The flow from this breather may be utilised to oil the rear chain. However if this supply becomes excessive, this is indicative of engine damage which requires urgent attention.

17.1 Crankcase breather may be taken from positions as shown. (A) 750cc (B) 650cc, 600cc and 500cc

On late models the breather vents to a froth tower on the oil tank and thence to the chainguard.

18 Oil filters - location and cleaning

1   All models have a gauze type filter incorporated in the main oil feed pipe from the oil tank. When the oil tank is drained for an oil change, this filter should be removed, washed in petrol and dried, before replacement. It forms an integral part of the nut of the oil pipe banjo union.

2   There is also a sump filter in the base of the left hand crankcase, identified by the very large bolt which forms the housing for the filter element. The bolt should be unscrewed and the gauze filter removed for cleaning with petrol by releasing the circlip and washer which precedes it.

19 Fault diagnosis

Symptom	Cause	Remedy
Engine 'fades' and eventually stops	Blocked air hole in filler cap	Clean
	Blockage in fuel system	Check and clean jets and filters
	Lack of petrol	Refill
Engine difficult to start	Carburettor flooding	Dismantle and clean carburettor. Check for punctured float *
Engine runs badly.. Black smoke from exhausts	Carburettor(s) flooding	Dismantle and clean carburettor. Check for punctured float *
Engine difficult to start. Fires only occasionally and spits back through carburettors	Weak mixture	Check for fuel in float chambers and whether air slides working
Oil does not return to oil tank	Damaged oil pump or oil pump drive	Stop engine immediately. Check oil tank contents. Remove timing cover to examine oil pump drive and seals
Engine joints leak oil badly	Pressure release valve inoperative	Dismantle valve and clean
Crankcase floods with oil when machine is left standing	Worn oil pump	Dismantle oil pump and eliminate end float of gears
Excess oil ejected from crankcase breather	Cylinders in need of rebore Broken or damaged piston rings	Rebore cylinders and fit oversize pistons Dismantle engine and renew rings

* or damaged float needle
and/or seating

Chapter 5 Ignition system

Contents

Specifications

Magneto (General Pre 1966)

Manufacturer	Lucas
Type	K2F
Breaker gap	0.012-0.015 in.
Contact breaker type	492854

Coil Lgnition (General Post 1966)

Coil Ignition (General Post 1966)

Ignition Coil

Manufacturer	Lucas
Type	12 volt

Distributor

Manufacturer	Lucas
Type	18D2

Contact Breaker

Manufacturer	Lucas
Type	40589
Gap	0.015 in fully open (0.35-0.40 mm)

Capacitor

Capacity	0.14-0.20 microfarads

Ignition Timing - B.T.D.C.

Model 88	30° Fully Advanced 6° Fully Retarded
All other models	32° Fully Advanced (28° on Atlas with 9:1 pistons) 8° Fully Retarded

Spark Plugs

	Champion	Lodge	KLG	NGK	Heat Range
Manufacturer					
Type (88, 99, 650 750)	N5	2HLN	FE75	B6ES	Hot
'SS' Models/High compression engines	N4	3HLN	FE80/FE100	B7ES	Cold

Size	14 mm.
Reach	¾ in.
Gap	
Magneto	0.020 in. (0.50 mm.)
Coil ignition	0.020 - 0.022 in. (0.50 - 0.55 mm.)

1 General Description

The spark necessary to ignite the petrol/air mixture in the combustion chambers is derived either from a battery and a single ignition coil or directly from a magneto. In general, the 1957 - 62 '88' and '99' models and '650' standard and de luxe models are equipped with coil ignition; all 'SS' models and 750s have a magneto pre 1966 and coil ignition post 1966. In both cases, the ignition unit is rotated at half engine speed; during each revolution it fires twice, providing a spark to each cylinder alternately.

In the 1957/8 period, Nortons started replacing the dynamo lighting system by an alternator unit mounted on the drive side end of the crankshaft. Initially this unit was used to generate alternating current which was rectified and used to charge a 6 volt battery. Further development occurred to provide coil igniton on some models and in the 1963/4 period the lighting system was uprated with a 12 volt battery whose charging is controlled via a Zener diode.

The distributor or the magneto is driven off the camshaft drive within the timing chest and is mounted on the rear 'ear' of the offside crankcase behind the engine. The battery is situated in the left hand case above the gearbox with the rectifier (if fitted) close by; either bolted to the battery case or to the tool tray under the seat (it requires a small amount of ventilation so should not be enclosed).

The Zenor diode (12 volt system only) is mounted in a large heat sink and is normally clamped to a bracket from the front left-hand tank fixing bolt. All models are equipped with an ammeter which is mounted in the headlamp nacelle, along with the ignition/lighting switch. Later models have the ignition switch by the oil tank and the lighting switch on the speedometer/rev counter bracket.

2 Magneto-Checking

1 The main advantage of the magneto over a coil ignition system, from the owners viewpoint, is the simplicity of checking the system in the event of ignition failure.
2 If the engine stops or will not start, commence with the premise that the condenser in the magneto has failed. Take out the plugs and ensure they are in good condition and correctly gapped. Kick over the engine and ensure the plugs are not sparking - if this is so, the magneto requires a complete overhaul by an electrical expert.
3 Unscrew the plug suppressor caps from the high tension leads and ensure that the screw connection is not loose nor covered in water and/or oil. Replace and re-check.
4 Remove the high-tension pick-ups from the side of the magneto by releasing the swinging spring clips, (early models) or the fixing screws (late models).
5 Unscrew the high tension leads from the pick-ups and ensure that the end of the high-tension lead is making contact with the fixed carbon pick-up. Clean any oil/water from the pick-up well. Replace and re-check; do not forget to replace the rubber cover over the screw connection to prevent the ingress of liquids (if the rubber has rotted, plastic insulation tape should be used to make a temporary seal).
6 Inspect the carbon brushes which are inserted into the pick-up and project into the magneto body. These are spring mounted and should be free to move and of sufficient length to reach the slip ring on the rotating armature. (i.e. renew the carbon brush if its length is 1/8 - 1/4 ins long). Clean the brush contact and the

slip ring with a clean rag - turn the magneto slowly to clean all around and to inspect the slip ring for damage. Slight damage may be removed with very fine emery; any serious damage will necessitate the removal of the magneto and possibly its exchange. Replace the pick-ups and recheck.

7 Turning to the contact breaker end of the magneto, disconnect the cut-out button screw to ensure this is not shorting. Re-test.
8 Remove the outer cover and ensure the contact points are clean, gapped and opening and closing correctly (see Section 4).

3 Ignition coils - checking

1 The ignition coil is a sealed unit, designed to give long service without attention. It is normally mounted near the petrol tank mountings or suspended from the engine steady: it also requires ventilation and should not be enclosed. If a weak spark and difficult starting cause the coil to be suspect, it should be checked by a Norton agent or an auto-electrical expert who will have the appropriate test equipment. A faulty coil must be renewed; it is not possible to effect a satisfactory repair.
2 If the complete ignition circuit fails, it is highly probable that the source of the fault will be found elsewhere.
3 A failed condenser or a dirty contact breaker assembly can give the impression of a faulty ignition coil and these components should be checked first before the coil itself receives attention.

4 Contact breakers - adjustments

Note: In the following test (M) indicates Magneto applications (C) indicates coil ignition applications.
1 To gain access to the contact breaker assembly, either remove the central swinging spring clip or remove the screw cap at the left hand end (M) and pull off the end cap complete with the cut-out wire; or release the side clips (2 off) and pull off the distributor cap complete with high tension leads (C). Later coil ignition models do not have the high tension leads on the contact breaker housing.

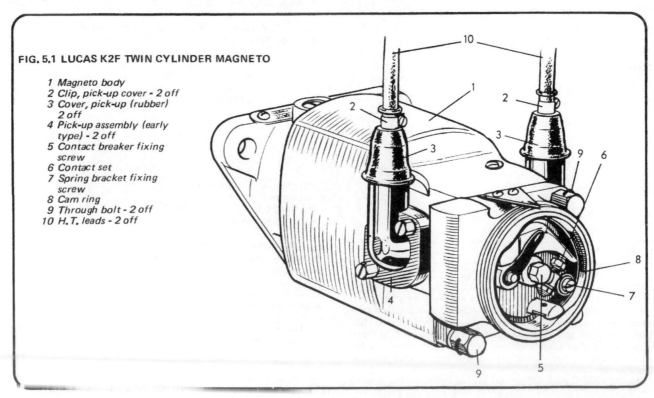

FIG. 5.1 LUCAS K2F TWIN CYLINDER MAGNETO

1 Magneto body
2 Clip, pick-up cover - 2 off
3 Cover, pick-up (rubber) 2 off
4 Pick-up assembly (early type) - 2 off
5 Contact breaker fixing screw
6 Contact set
7 Spring bracket fixing screw
8 Cam ring
9 Through bolt - 2 off
10 H.T. leads - 2 off

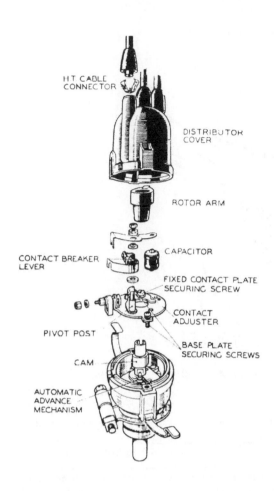

HT CABLE CONNECTOR

DISTRIBUTOR COVER

ROTOR ARM

CONTACT BREAKER LEVER

CAPACITOR

FIXED CONTACT PLATE SECURING SCREW

CONTACT ADJUSTER

PIVOT POST

BASE PLATE SECURING SCREWS

CAM

AUTOMATIC ADVANCE MECHANISM

FIG. 5.2 LUCAS 18D2 DISTRIBUTOR

2 Rotate the engine slowly by means of the kickstarter until the contact breaker points are in the fully open position. Examine the faces of the contacts; if they are pitted or burnt, it will be necessary to remove them for further attention, as described in Section 5 of this Chapter. In order to obtain a full view of the surfaces, gently lift the sprung contact with a screwdriver.

3 To adjust the gap, keep the engine in the position giving maximum gap and:- (c) slacken the screw securing the fixed contact plate. Insert a screwdriver between the two studs or pips on the base plate and the notch in the fixed contact plate. Adjust the position of the plate until the gap is 0.015in. Tighten the securing screw and recheck the gap. On earlier models, fitted with a magneto, slacken the locking nut of the adjustable fixed contact point and adjust the gap to 0.012 - 0.015. A magneto spanner is required for this operation. Tighten the locking nut (do not overtighten) and recheck the gap. On later models, release the fixed contact plate securing screw and rotate the points assembly using a screwdriver as shown in fig 5.3

In all cases, recheck the gap on the other cylinder and reach the best compromise if a discrepancy exists by "sharing" the error between both cylinders.

4 It is important to ensure that the gap is re-set when the points are fully open, otherwise a false reading will result.

5 Before replacing the contact breaker cover, smear the contact breaker cam sparingly with grease and add a few drops of engine oil to the lubricating felt. The cover plate must be replaced with the small drain/breather hole facing downwards, to prevent the ingress of water.

5 Contact breaker points - removal, renovation and replacement

1 If the contact breaker points are pitted or burned, they should be removed for dressing. Badly worn points should be renewed without question, as should points which will require a substantial amount of material to be removed before their faces can be restored.

2 To remove the contact breaker points, remove:

Late Magneto and all coil ignition models
The nut securing the spring to the terminal post; pull off the moving contact; unscrew the fixed contact plate securing

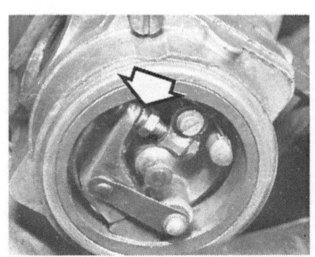

4.3a Early type magneto points assembly - hexagon screw adjuster arrowed

4.3b Check the gap with a feeler gauge

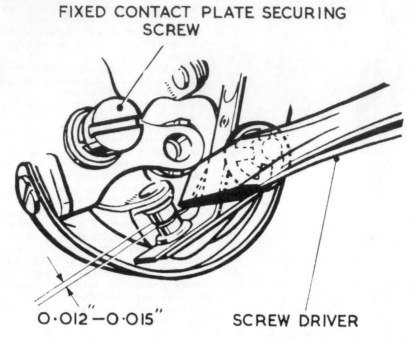

FIXED CONTACT PLATE SECURING SCREW

0·012″ – 0·015″

SCREW DRIVER

FIG. 5.3 ADJUSTING THE CONTACT BREAKER POINTS (LUCAS MAGNETO LATE TYPE)

screw and pull off the plate.

Early magneto ignition models

The swinging spring clip from the moving contact; pull off the moving contact; unlock and unscrew the fixed contact until it is removable.

3   On coil ignition and the later magneto fitted machines make special note of the way in which the insulators and the fibre washers are arranged especially around the moving contact spring location.

4   The points should be dressed with an oilstone or fine emery cloth. Keep them absolutely square during the dressing operation, otherwise they will make angular contact with each other when reassembled and will quickly burn away.

5   Clean the points with petrol to remove all traces of abrasive and oil, then reassemble them by reversing the dismantling procedure. Make sure the insulating washers and collar are replaced in the correct order, where fitted, and that the pivot of the moving contact rocker arm has a light smearing of grease

6   When reassembly is complete, check and if necessary, re-set the contact breaker gap.

6   Auto-advance assembly - examination

1   The auto-advance assembly is located either in the timing chest on the end of the distributor/magneto shaft (and is part of the chainwheel assembly, or is located within the distributor body (C). Access in the former case is only possible after removing the timing cover (Chapter 7 section 10).

2   The auto-advance assembly comprises two spring-loaded weights which fly apart with centrifugal action. They are connected to the contact breaker cam so that as the engine speed rises, the cam moves independently of the camshaft from which the assembly is driven, thereby advancing the ignition timing.

3   Visual inspection will show whether tired or broken springs are restricting the freedom of movement, or whether rust, resulting from condensation, is preventing the weights from moving smoothly. The only maintenance required is an occasional drop

of oil on the various pivots to ensure they do not run dry.

7   Condensers - removal and testing (coil ignition only)

1   The condenser is included in the contact breaker circuitry to prevent arcing across the contact breaker points at the moment of separation. It is connected in parallel with its own set of points and if the condenser should fail, an ignition fault will occur in the circuit involved. The condenser is retained by a spring plate which may require slackening before it may be removed.

2   If the engine is difficult to start or if a persistent misfire occurs it is possible that the condenser has failed. To check, separate the contact breaker points of the cylinder concerned by hand, with the ignition switched on. If a spark occurs across the points and they have a blackened and burnt appearance, the condenser can be regarded as unserviceable.

3   It is not possible to test the condenser without the appropriate test equipment. It is best to fit a new replacement in view of the low cost involved and observe the effect on engine performance.

8   Cut-out switch - Magneto models only

1   In order to stop the engine, a spring contact in the contact breaker cap can be shorted, to earth the ignition system.

2   The cut-out button is mounted on the handlebars. It should be checked regularly to ensure it is in good condition, or it may short the ignition involuntarily, either permanently or spasmodically, causing the engine to misfire or stop altogether stop altogether.

9   Ignition Timing

1   The only exact method of setting the ignition requires the use of a timing disc. This is a cheap item which should be purchased or a replica made.

2 In order to time the ignition using the timing disc, it is necessary to remove the primary chaincase and attach the disc to the end of the crankshaft. A piece of wire under one of the barrel retaining bolts should be used as a pointer.

3 For magneto equipped models, the timing cover must be removed for all timing adjustments. On coil ignition models, it is only necessary to remove the timing cover when initially setting the ignition.

4 Ensure that the distributor/magneto driving chain is correctly adjusted ie. it should have 3/16 in up and down movement at its tightest point. The adjustment is made by slackening the top two Allen screws/nuts and tapping the ignition unit as necessary with a hide hammer to pivot it about the bottom stud.

5 Ensure that the contact breaker gap is correctly set to 0.015 ins (Distributor) or 0.012ins (Magneto).

6 Turn the motor to top dead centre (TDC) and tighten the crankshaft nut to retain the disc after aligning the wire pointer to the O^o mark.

7 Remove the sparking plugs and check that the marked tooth of the camshaft sprocket is in the 11 o'clock position. This position denotes TDC on the firing stroke of the drive side cylinder.

Magneto Ignition Models

8 Slacken the bolt retaining the automatic advance unit device. This bolt has a double start thread which acts as a self-extractor for the unit. Unscrew the bolt until the A.T.D. unit is free from the magneto shaft. It will slacken initially then, tighten again as it commences to draw the unit off the magneto shaft.

9 Using a small section of plug lead offcut, prop the automatic advance unit into the fully open position. (ie. fully advanced).

10 Rotate the engine backwards until the pointer on the timing disc indicates the correct amount of degrees advance. (See specifications).

11 Rotate the contact breaker "forward" until the points just commence to open. This position should be accurately determined using either a thin strip of cigarette paper or a small bulb and battery connected from the moving contact to earth.

12 Hold the shaft steady and screw in the auto-advance unit bolt finger tight.

13 Remove the wedge and tighten the bolt fully. Replace the wedge and recheck the timing. It must be set very accurately hence the need for a timing disc which must first be exactly at zero when the engine is at top dead centre.

9.9 Use of a stud to keep the auto-advance unit in the fully advanced position

Coil ignition models

14 Remove the roll pin and sprocket from the end of the distributor shaft using a sprocket extractor or judicious use of a lever.

15 Rotate the distributor so that the brass contact of the rotor is pointing directly downwards. Check that the roll pin holes in the sprocket and spindle line-up. If they do not, ease the sprocket off the spindle and remesh it with the chain to the nearest tooth. Replace on the spindle and re-fit the roll pin. Note:- when removing (using a small diameter drift) or replacing the roll pin, a second hammer head should be placed on the shaft in order to provide some support.

16 Remove the distributor cap and rotor and the contact breaker base plate. Proceed as in paragraph 9.

17 See paragraph 10

18 Temporarily replace the contact breaker plate and slacken the pinch bolt on the distributor clamp. Rotate the distributor body until the points just begin to open - see paragraph 11. Note: On coil ignition models the moment of separation may be determined by switching on the ignition and connecting a voltmeter from the moving contact to earth. When the points are closed the reading will be zero; when the points separate the voltmeter will show a reading.

19 Tighten the clamping bolt and recheck the setting.

20 Remove the automatic advance wedge and replace the contact breaker plate. Recheck the ignition by "advancing" the rotor in turn the rotor to the fully advanced position.

21 Replace the timing cover.

22 Replace the sparking plugs and plug leads, also the magneto end cap/distributor cap. The rearmost high tension lead should be attached to the left hand sparking plug.

23 Attempt to start the engine. If it will not start and only minor adjustments to the ignition timing have been made, then it is probable that the ignition was not previously set up as above. In this case, change over the plug leads. If the engine still will not start and a check shows a healthy spark, look for the cause in the carburettor system.

10 Spark plugs - checking and resetting the gaps

1 14 mm spark plugs with a ¾ inch reach are fitted to all Norton twin models. Refer to the Specifications section of this Chapter for the list of recommended grades. Always use the grade recommended, or the direct equivalent in another manufacturer's range.

2 Check the gap at the plug points every 3000 miles. To reset the gap, bend the outer electrode closer to the central electrode and check that the correct feeler gauge can be inserted. Never bend the central electrode or the insulator will crack, causing engine damage if particles fall in whilst the engine is running. Do not lever the outer electrode with a screwdriver between it and the centre electrode.

3 The condition of the spark plug electrodes and insulator can be used as a reliable guide to engine operating conditions, with some experience. See accompanying illustrations.

4 Always carry at least one spare spark plug of the correct grade. This will serve as a get-you-home means if one of the spark plugs in the machine should fail.

5 Never overtighten a spark plug, otherwise there is risk of stripping the threads from the cylinder head, especially in the case of one cast in light alloy. A stripped thread can be repaired by using what is known as a 'Helicoil' thread insert, a low cost service of cylinder head reclamation that is operated by many dealers.

6 Use a spark plug spanner that is a good fit, otherwise the spanner may slip and break the insulator. The plug should be tightened sufficiently to seat firmly on its sealing washer.

7 Make sure the plug insulating caps are a good fit and free from cracks. The caps contain the suppressors which eliminate radio and TV interference; in rare cases the suppressors have

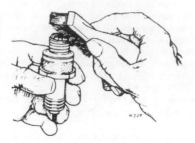

Cleaning deposits from electrodes and surrounding area using a fine wire brush

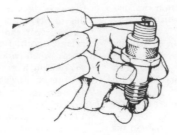

Checking plug gap with feeler gauges

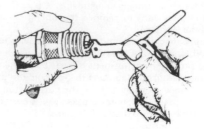

Altering the plug gap. Note use of correct tool

FIG. 5.4a SPARKING PLUG MAINTENANCE

White deposits and damaged porcelain insulation indicating overheating

Broken porcelain insulation due to bent central electrode

Electrodes burnt away due to wrong heat value or chronic pre-ignition (pinking)

Excessive black deposits caused by over-rich mixture or wrong heat value

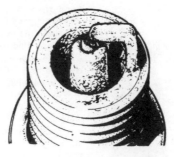

Mild white deposits and electrode burnt indicating too weak a fuel mixture

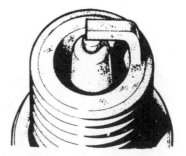

Plug in sound condition with light greyish brown deposits

FIG. 5.4b SPARKING PLUG ELECTRODE CONDITIONS

developed a very high resistance as they have aged, cutting down the spark intensity and giving rise to ignition problems.

8 The spark plugs should be replaced after 6000 miles on a machine which is driven hard. Longer periods of life will be achieved under touring conditions but the plugs should be replaced when the plug electrodes become rounded and the side electrode had a significant droop to maintain the gap.

9 Plugs may be cleaned by sand blasting and/or washing in petrol with a stiff, wire brush.

10 If the machine has been tuned or is ridden extremely hard, a change in spark plug grade may be necessary from that specified because the standard plug may run too hot and burn away. In this case, the plugs should be replaced with a colder running type (ie. a Champion N5 should be replaced by an N4). Conversely if an 'SS' model is continually used in town traffic, the plugs may "oil up" because they are running too cold and cause misfiring. The cold plug should then be replaced by a hotter plug.

In either case, if changing the plugs does not remove the malfunction, replace the standard plugs and look elsewhere for the problem - eg. carburation and ignition systems.

11 Fault diagnosis

Symptom	Cause	Remedy
Engine will not start	Water or condensation in magneto/distributor.	Dry off
	Break or short circuit in electrical system	Switch off (c) and check wiring.
	Contact breaker points closed	Re-adjust contact breaker points.
Engine misfires	Faulty condenser (c)	Replace condenser and re-check
	Incorrect ignition timing	Check accuracy of ignition timing of cylinder involved.
	Fouled or incorrect grade of spark plug	Clean plug and/or replace with correct grade
	Break or short circuit in electrical system	Check wiring.
Engine lacks response and overheats	Reduced contact breaker gaps	Check and reset
	Jammed auto-advance unit	Check whether balance weights are free to move.
Engine 'fades' when under heavy load	Pre-ignition	Check the ignition timing
		Replace plugs, using only recommended grades.
		Decarbonise the cylinder head
Engine will not start in the emergency position (C)	Short or break in electrical system	Check wiring.
	Low output from alternator	Check output

(c) indicates coil ignition models only.

Chapter 6 Frame and forks

Contents

Specifications

Forks

Type	Telescopic with hydraulic damping and internal springs. (External springs on early "Roadholder" forks and some racing machines).
Oil capacity	150cc (5 fl. oz) each leg 170cc Scrambles models

Rear Suspension Units

Manufacturer	Girling or Armstrong (1956)

Frame
Featherbed Type

Dimensions		ins	cms
Wheelbase		55½	141
Length (overall)		84	213
Width	"	25½	64½
Ground Clearance		6¼	16
Weight (Dry)		398 lb	180½ kg

1 General Description

Two types of frame are used to house the twin cylinder engine; the "featherbed" frame and the "scrambles" frame.

Featherbed frame

This frame is a development of that used to carry the Manx Norton motor, first introduced to racing during the 1950 season. Since that time it has developed into a completely brazed construction in place of the original bolted -on rear end. Duplex frame tubes completely encircle the motor; braced laterally at four points an extremely rigid construction is obtained. Originally the top frame tubes were extremely wide to enable the removal of the Manx cylinder head. However, in 1960, the top frame tubes were waisted to produce the "slimline featherbed". This enabled a narrower tank to be fitted and improved rider comfort.

The 'featherbed' frame provides the Dominator engines of all capacities with what is probably the best production machine handling characteristics ever achieved. In addition, the frame is virtually maintenance free.

Scrambles frame

Introduced to supplement the AMC range and provide a higher power to weight ratio, the light, scrambles type frame is used only to carry the 750cc unit under both Norton and Matchless banners. It is of more conventional design with a single top tube and duplex front downtubes to carry the engine.

Both frames incorporate versions of the famous "Roadholders" forks and utilise similar wheel hubs, brakes, swinging arm and rear suspension systems.

Note:- the main differences amongst the Roadholder fork versions is around the top of the fork legs. Early versions have seal covers held by two 4BA screws. External spring versions have a short, threaded seal retainer. If this is omitted the whole front fork sliders and wheel etc will fall out if the wheel leaves the ground.

2 Front forks - removal from frame

1 It is unlikely that the front forks will need to be removed from the frame as a complete unit, unless the steering head bearings require attention or the forks are damaged in an accident.

2 Position the machine firmly on a wooden box or block placed under the lower frame tubes and remove the front wheel. The box should be of sufficient height so that the forks are fully extended and the front wheel is clear of the ground.

3 Disconnect the front brake cable at the brake operating arm. It is secured to the arm by a spring clip and clevis pin. Unscrew the cable adjuster from the brake plate.

4 Slacken the pinch bolt in the end of the left hand fork leg and remove the nut from the right hand end of the spindle. Insert a tommy bar through the left hand end of the spindle so that the spindle can be withdrawn whilst the front wheel is supported by hand. The wheel can now be withdrawn from the forks, after the brake plate anchor has disengaged from the abutment on the lower right hand fork leg.

5 Slacken the nut at the base of each bolt in the top of each fork leg, so that the fork damper rod (late models only) can be unscrewed and the bolt freed. Unscrew the speedometer cable (and tachometer cable where fitted) and allow them to hang. Remove the two bolts securing the headlamp and leave the headlamp suspended.

6 Remove the chromium plated dome nut in the centre of the fork top yoke, together with its washer. The top yoke can then be driven off the steering head stem and stanchion tapers by hitting the underside of the yoke with a rawhide mallet. Remove the nut and dust cover from the stem and then drive the stem downwards through the steering head bearings, to release the forks as a complete unit, complete with mudguard. The latter is detached by removing the two nuts on the inside of each fork leg and the bolts which anchor the stays to the lower end of each fork leg. Note:- As the head is driven down, the lower ball bearing cups will separate and the balls drop out unless care is taken either to wrap a rag around the bearing or the balls are removed as soon as the gap is large enough.

7 Where clip-on handlebars are fitted, it will also be necessary to disconnect all cables and wires to the frame eg. clutch cable, magneto earth, unless the clamp bolts are removed and the handlebars are slid upwards, off their respective fork legs.

8 Removal of individual fork legs only (late type Roadholders). Proceed as above to include paragraph 5 although there is no need to disconnect the drive cables or the headlamp. Also refer to paragraph 7. The clip-ons and the headlight attachments must be freed from the fork leg. Free the mudguard, as described in paragraph 6.

9 Slacken the pinch bolts in the lower yoke. Replace the chromium plated bolts in the top of each fork leg so that at least six threads engage. With a block of wood interposed between the bolt head and the hammer, drive the bolt downwards in order to break the taper joint of the fork stanchions. Remove the bolts again. Unless the fork leg is corroded to the lower yoke, it will fall to the floor once the taper is broken and it should be held to prevent this happening. Note:- if rubber gaiters are fitted as on the scrambler models in place of the steel tubes of the featherbed model, they should be freed before the fork leg is removed.

FIG. 6.1 FRONT FORK ASSEMBLY (ROADHOLDER FORKS)

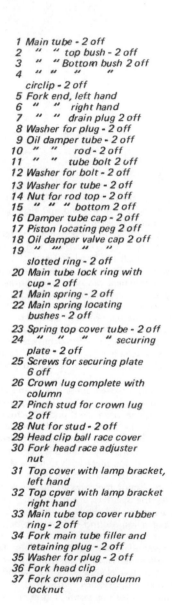
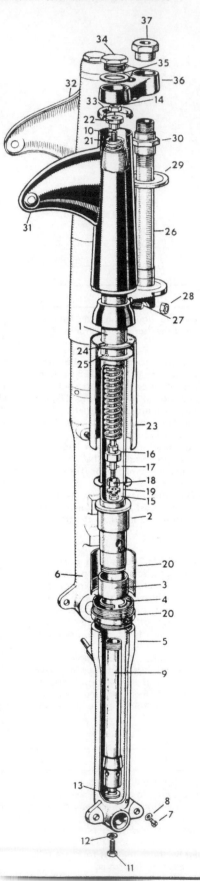

1 Main tube - 2 off
2 " " top bush - 2 off
3 " " Bottom bush 2 off
4 " " " "
  circlip - 2 off
5 Fork end, left hand
6 " " right hand
7 " " drain plug 2 off
8 Washer for plug - 2 off
9 Oil damper tube - 2 off
10 " " rod - 2 off
11 " " tube bolt 2 off
12 Washer for bolt - 2 off
13 Washer for tube - 2 off
14 Nut for rod top - 2 off
15 " " " bottom 2 off
16 Damper tube cap - 2 off
17 Piston locating peg 2 off
18 Oil damper valve cap 2 off
19 " " " "
   slotted ring - 2 off
20 Main tube lock ring with
   cup - 2 off
21 Main spring - 2 off
22 Main spring locating
   bushes - 2 off
23 Spring top cover tube - 2 off
24 " " " " securing
   plate - 2 off
25 Screws for securing plate
   6 off
26 Crown lug complete with
   column
27 Pinch stud for crown lug
   2 off
28 Nut for stud - 2 off
29 Head clip ball race cover
30 Fork head race adjuster
   nut
31 Top cover with lamp bracket,
   left hand
32 Top cpver with lamp bracket
   right hand
33 Main tube top cover rubber
   ring - 2 off
34 Fork main tube filler and
   retaining plug - 2 off
35 Washer for plug - 2 off
36 Fork head clip
37 Fork crown and column
   locknut

2.3 Disconnect brake cable at the operating arm (B) Note: Drain plug screws (A)

2.4 Pinch bolt (A) retains front wheel spindle on left-hand side

2.5 Large chromium headed bolts retain stanchions to upper fork yoke

2.9 Slacken pinch bolts in lower fork yoke to free individual fork legs

3 Front forks - dismantling the fork legs

1 Drain the fork leg by inverting it over a suitable receptacle or by removing the drain plug just above the wheel spindle orifice in the lower fork leg. Unscrew the locknut at the upper end of the damper rod and withdraw the locating bush and fork spring.

2 Clamp the lower fork leg in a vice, using aluminium clamps to prevent marking the outer surface. The fork leg should be clamped in the horizontal position.

3 Unscrew the chromium plated threaded collar (later models only). The collar, which has a right hand thread, may be tight, necessitating the application of a strap spanner or heat. When the collar is unscrewed fully, grasp the fork stanchion with both hands and with a number of sharp upward pulls, displace the oil seal, paper washer and top bush. The stanchion can now be lifted away from the lower fork leg completely.

4 Remove the damper tube anchor bolt which is recessed into the end of the lower fork leg, within the portion that carries the wheel spindle. Do not lose the fibre washer on which the damper tube seats. The damper rod with damper unit attached can then be lifted out of the lower fork leg as a unit. Do not use a rod or a Phillips screwdriver through the holes in the bottom. Secure the

damper tube carefully but firmly in a vice (it is easily crushed) and unscrew the alloy damper tube cap. This will free the damper rod. There is no necessity to dismantle the damper rod assembly further, unless the damper valve is to be renewed. Use two damper rod nuts tightened together as locknuts to prevent the rod turning so that the damper rod locknut and square washer can be removed to free the damper valve.

5 To gain access to the lower fork bush, remove the square section circlip from the bottom of the fork stanchion (later models only).

G15CS MODELS

The forks on the scrambles models are very similar to the featherbed framed models. The main differences include 2" longer fork stanchions, EXTERNAL fork springs and rubber gaiters. The upper fork bush and seal are retained by a threaded collar.

4 Front forks - general examination

1 Apart from the oil seals and bushes, it is unlikely that the forks will require any additional attention, unless the forks

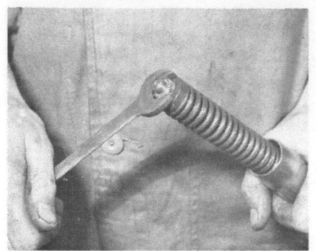

3.1 Unscrew damper rod locknut and withdraw bush and fork spring

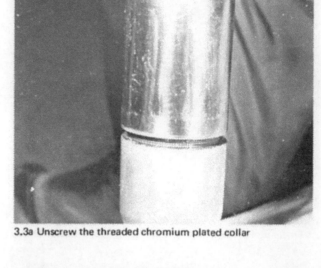

3.3a Unscrew the threaded chromium plated collar

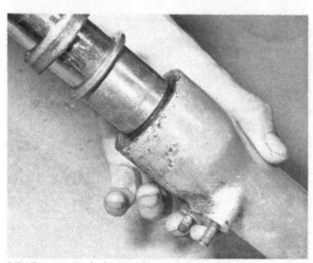

3.3b Remove the fork stanchion complete with oil seal and sliding bush

3.4a The damper tube retaining bolt and washer are recessed into lower end of fork leg

3.4b Damper tube seats on fibre washer

springs have weakened or the damper units have lost their efficiency. If the fork legs or yokes have been damaged in an accident, it is preferable to renew them. Repairs are seldom practicable without the appropriate repair equipment and jigs. Furthermore, there is also risk of fatigue failure.

2 Visual examination will show whether either the fork legs or the yokes are bent or distorted. The best check for the fork stanchions is to remove the fork bushes, as described in paragraphs 3 and 5 of the preceding Section, and roll the stanchions on a sheet of plate glass, or on a surface plate. Any deviation from parallel will immediately be obvious.

5 Front forks - renewal of oil seals

1 If the fork legs have shown a tendency to leak oil or if there is any other reason to suspect the efficiency of the oil seals, they should be renewed without question.

2 The oil seals are retained in the cupped portion of the lower fork legs by the screwed collar. They are displaced by following the procedure detailed in Section 3.3 of this Chapter. The new oil seal should be tapped into position with a tubular drift of the correct size, after the top fork bush and washer have been located in the lower fork leg first. Note: the fork seals should be fitted with the exposed wire coil downwards.

6 Front forks - examination and renewal of bushes

1 Some indication of the amount of wear in the fork bushes can be gained when the forks are being dismantled. Pull each fork stanchion out of the lower fork leg until it reaches the limit of its extension and check the side play. In this position the two fork bushes are closest together, which will show the amount of play to its maximum. Only a small amount of play which is just perceptible can be tolerated. If the play is greated than this, the bushes are due for replacement. Check also the slider for wear, which usually occurs half way up.

2 It is possible to check for play in the bushes whilst the forks are still attached to the machine. If the front wheel is gripped between the knees and the handlebars rocked to and fro, the amount of wear will be magnified by the leverage at the handle-bar ends. Cross-check by applying the front brake and pushing and pulling the rear wheel backwards and forwards. It is important not to confuse any play which is evident with slackness in the steering head bearings, which should be taken up first, or even with play at the brake plate stop on the right hand fork leg.

3 The fork bushes can be removed from the stanchion when the circlip at the lower end of the stanchion has been detached.

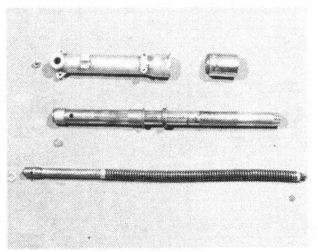

6.1 The fork leg, completely dismantled

7 Steering head bearings - examination and renewal

1 Before commencing to reassemble the forks, the steering head bearing should be inspected. Most machines have ball bearings running in split cups at the top and bottom of the yoke. This type of arrangement is adjustable for wear. Note:- conversions are available to replace the cups and balls with taper roller bearings.

2 One cup of each bearing is a drive fit in the top bottom of the steering head and should not be removed unless pitting has damaged the rolling surface. Pitting is caused by running with the steering head bearings slack.

3 Replace all the ball bearings if any show signs of pitting or corrosion. This also applies to the outer ball cups.

4 The lower, outer ball cup is a drive fit on the lower yoke. A tapered drift or screwdriver is required to remove it.

5 The upper, outer ball cup is located by the large adjusting nut. The steering head, on replacement, should be devoid of 'fore and aft' movement and free from any frictional resistance. Note: the adjusting nut is retained in position by the upper fork and the domed nut. When adjusting the steering head with the forks fully assembled, the domed nut, the upper large chromium fork nuts and the fork leg pinch bolts on the lower yoke should all be slackened. When the adjustment has been made, retighten

all these bolts and recheck the setting, levering from the end of the fork legs (refer to Section 8).

8 Front forks - reassembly

1 To reassemble the forks, follow the dismantling procedure in reverse. Particular care is necessary when refitting the oil seals or new replacements. Grease both the inside of the seal and the fork stanchion to obviate risk of damage to the feather edge of the seal.

2 Tighten the steering head carefully, so that all play is eliminated without undue stress on the bearings. The adjustment is correct if all play is eliminated and the handlebars will swing to full lock on their own accord, when given a light push on one end.

3 It is possible to place several tons pressure quite unwittingly on steering head bearings of the adjustable type, if they are overtightened. The symptom of overtight bearings is a tendency for the machine to roll at low speeds or the handlebars to oscillate, even though the handlebars may appear to turn quite freely when the machine is on the centre stand.

4 Difficulty may be encountered due to the reluctance of the stanchions to seat correctly on their tapers within the top fork yoke. The chromium plated bolts can be used to pull the stanchions into position, if necessary, without passing through the large washers as a temporary expedient. Do not use force unless a good depth of thread has engaged with the stanchion.

5 Tighten the chromium plated top bolts first, then the steering head stem nut, followed by the pinch bolts throught the lower fork yoke. If the forks are incorrectly aligned or stiff in action, slacken each of them a little and bounce the forks several times before retightening from the top downwards, finishing with the pinch bolts. This same technique can be used if the forks are misaligned after an accident. Often the legs will twist within the fork yokes, giving the impression of more serious damage, even though none has occurred.

6 Do not omit to slacken and raise the two chromium plated fork bolts so that the correct amount of damping fluid can be added to each fork leg. Check that the drain plugs in each fork leg have been replaced first! The bolts should be tightened to a torque setting of 40 ft lb.

9 Front forks - damping action

1 Each fork leg contains a pre-determined quantity of damping fluid, used to control the action of the compression springs within the forks when various road shocks are encountered. If the damping fluid is absent, or the damper units ineffective, there is no control over the rebound action of the fork springs and fork movement would become excessive, giving a very 'live-ly' ride. Damping action restricts fork movement and is progressive in action; the effect becomes more pronounced as the rate of deflection increases. The Norton fork is particularly good in this respect, since the two-way damping system applies to both compression and extension.

2 Reference to the accompanying illustration will show how the Norton system operates. When the forks are compressed the stanchion tube and damper rod remain stationary as the lower fork leg and damper tube rises. As the damper tube rises, the valve of the end of the damper rod is raised, permitting oil to pass the seating washer. Because the displaced oil cannot escape from the sealed end of the damper tube, excess oil is expelled through the bleed holes in the lower end of the fork stanchion, aided by the vacuum which is produced as the gap between the upper and lower fork brushes widens. It is the restrictive action of the bleeder holes which slows down the fork movement. When the lower fork leg approaches its limit of travel, a cone-shaped plug in the bottom of the leg commences

to enter the end of the fork stanchion tube, thereby progressively cutting off the alternative passageway for the oil.Oil is therefore compressed in the extreme bottom end of the lower fork leg to provide an effective bump stop, leaving only the bleed holes in the damper tube as a controlled exit to prevent a complete hydraulic lock. When the fork extends, the lower fork leg and the damper unit descend, leaving oil trapped between the upper and lower fork bushes. Because the damper rod is also descending, the damper valve is forced against its seating and is held closed. The pressure exerted allows oil to escape via the tiny clearance between the damper rod and the cap of the damper unit, into the area above. As the fork continues to extend, the oil to pass the seating washer. Because the displaced oil cannot between the fork bushes closed and is forced into the lower fork leg through a large hole in the stanchion and then a smaller hole in the damper body, which effectively slows down fork action in a progressive manner and prevents the forks from reaching the upper limit of travel.

10 Frame assembly - examination and renovation

1 If the machine is stripped for an overhaul, this affords an excellent opportunity to inspect the frame for signs of cracks or other damage which may have occurred during service, eg. the stand wearing through the bottom frame tube. Frame repairs are best entrusted to a frame repair specialist who will have all the necessary jigs and mandrels necessary to ensure correct alignment. This type of approach is recommended for minor repairs. If the machine has been damaged badly as the result of an accident and the frame is well out of alignment, it is advisable to renew the frame without question or, if the amount of money available is limited, to obtain a sound replacement from a breaker's yard.

2 If the front forks have been removed from the machine, it is comparatively simple to make a quick visual check of alignment by inserting a long tube that is a good push fit in the steering head races. Viewed from the front, the tube should line up exactly with the centre line of the frame. Any deviation from the true vertical position will immediately be obvious; the steering head is a particularly good guide to the correctness of alignment when front end damage has occurred. More accurate checking must be carried out when the frame is stripped.

11 Swinging arm rear suspension - dismantling, examination and renovation

1 Before access to the swinging arm fork is available, it is necessary to place the machine on the centre stand so that it is standing firmly on level ground. Remove the rear wheel complete with brake drum and sprocket. Refer to Chapter 7 for the recommended procedure relating to this task.

2 Remove the chainguard for ease of manipulation along with the final drive chain. Do not forget the chain oiler pipe.

3 Remove completely the bolts retaining the lower end of the rear suspension units to the rear fork. Knock the rubber mounting clear.

4 Only the large diameter spindle now attaches the swinging arm to the frame gusset plates. Remove the left hand spindle nut and tap out the spindle. Remove the fork. Note:- Most machines will have 'Clayflex' bushes in the swinging arm pivot although many owners have put in the "Manx" modification of phosphor-bronze bushes to provide a more rigid arrangement. In the former case, owing to the small amount of radial movement being taken up in the rubber mounting, the spindle and inner brush bearing corrode to form a solid unit. Hence, if the spindle does not easily tap out, replace the nut almost completely and try a few "hefty whacks". If the spindle still does not move, it is recommended that the machine is taken to a Norton agent for the rear fork to be removed.

5 The bushes which form the bearing surface for the pivot tube are a tight fit in the inner ends of the eyes of the swinging arm fork. If play is evident between the pivot pin and the bearings

11.2 Nuts and bolts (2 off) retain the chainguard

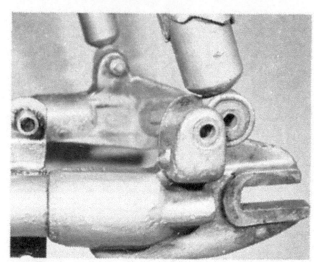
11.3 Knock out the suspension units after removing bolts

11.4 The swinging arm left hand spindle nut

AT REST (STATIC)   COMPRESSING   C1

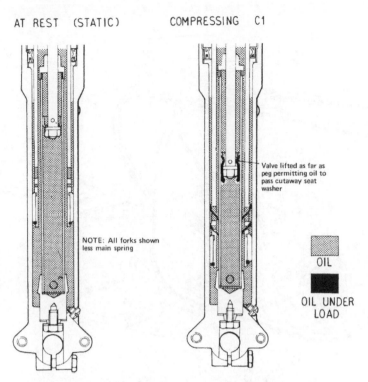

Valve lifted as far as peg permitting oil to pass cutaway seat washer

NOTE: All forks shown less main spring

OIL

OIL UNDER LOAD

Fig. 6.2a. Diagrammatic illustration of fork damping action (at rest and compressing)

FINAL COMPRESSION   C2   EXTENDING

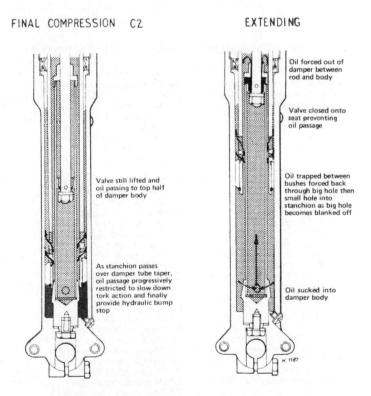

Valve still lifted and oil passing to top half of damper body

As stanchion passes over damper tube taper, oil passage progressively restricted to slow down fork action and finally provide hydraulic bump stop

Oil forced out of damper between rod and body

Valve closed onto seat preventing oil passage

Oil trapped between bushes forced back through big hole then small hole into stanchion as big hole becomes blanked off

Oil sucked into damper body

Fig. 6.2b. Diagrammatic illustration of fork damping action (Final compression and extending)

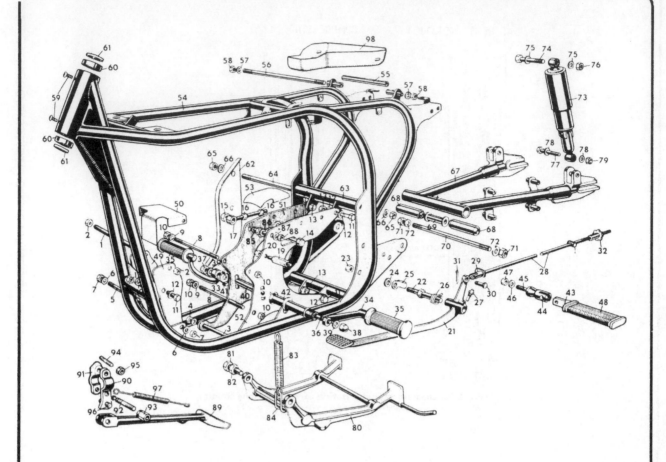

FIG. 6.3 FRAME AND AUXILIARIES SLIMLINE ('FEATHERBED' TYPE)

1 Crankcase to engine plate stud, front, top
2 Nut for stud - 2 off
3 Crankcase to engine plate bolt (rear, bottom)
4 Nut for bolt
5 Crankcase to engine plate and frame stud (bottom)
6 Washer for stud - 2 off
7 Nut for stud - 2 off
8 Crankcase to engine plate stud (rear), top and middle 2 off
9 Washer for stud - 4 off
10 Nut for stud - 4 off
11 Front and rear engine plate to frame bolt - 6 off
12 Washer for bolt - 12 off
13 Nut for bolt - 6 off
14 Gearbox top bolt
15 Nut for top bolt
16 Gearbox front chain adjuster nuts - 2 off
17 Front chain adjuster
18 Brake arm pin
19 Chaincase inner portion support stud
20 Nut for stud
21 Rear brake pedal
22 Rear brake pedal spindle
23 Nut for spindle
24 Distance piece for spindle

25 Stop for pedal
26 Return spring for pedal
27 Grease nipple and retaining pin for pedal
28 Brake rod
29 Brake rod jaw joint assembly
30 Pin for jaw joint
31 Split pin
32 Brake rod adjuster nut
33 Footrest hangar, right hand
34 " " left "
35 Footrest rubber - 2 off
36 " rod
37 Nut for rod, plain
38 " " " domed
39 Washer for rod - 2 off
40 Distance piece between engine plates
41 Footrest tube serrated end, right hand
42 Footrest tube serrated end, left hand
43 Pillion footrest bar - 2 off
44 " " lug - 2 off
45 Stud for pillion footrest and silencer fixing - 2 off
46 Pillion footrest lug fixing washers - 2 off
47 Pillion footrest lug fixing nut - 2 off
48 Pillion footrest rubbers 2 off

49 Front engine plate - 2 off
50 Front engine plate cover
51 Rear " " right hand
52 Rear engine plate left hand
53 Rear engine plate cover
54 Frame (only)
55 Frame distance tube (top rear)
56 Stud for distance tube
57 Washer for stud - 4 off
58 Nut for stud - 2 off
59 Screws for headlug hole blanking - 2 off
60 Frame headlug bearing (top and bottom) 2 off
61 Felt washer for bearing 2 off
62 Gusset plate cover
63 Cross tube, rear engine fixing
64 Rod for cross tube
65 Nut for rod - 2 off
66 Washer for rod - 2 off
67 Swinging arm only
68 Silent bloc bearing - 2 off
69 Spacer tube for bearing
70 Swinging arm rod
71 Nut for rod - 2 off
72 Washer for rod - 2 off
73 Rear shock absorber unit (solo) 2 off

74 Top fixing bolt for shock absorbers - 2 off
75 Washers for top fixing bolts 4 off
76 Nuts for top fixing bolts 2 off
77 Bottom fixing bolts for shock absorbers 2 off
78 Washers for bottom fixing bolts 4 off
79 Nuts for bottom fixing bolts 2 off
80 Centre stand
81 Bolt for stand 2 off
82 Nut for bolt - 2 off
83 Spring for stand
84 Clip for spring
85 Anchor stud for spring
86 Washer for stud (thick)
87 Washer for stud (thin)
88 Nut for stud
89 Side prop stand leg
90 Side prop stand clip lug front
91 Side prop stand clip lug rear
92 Clip lug and spring fixing stud
93 Nut for fixing stud
94 Clip lug stud
95 Nut for stud
96 Fulcrum pin
97 Prop stand return spring
98 Tool tray complete

the pivot pin and bearings should be renewed as a complete set. Note that a small amount of movement at the bearing will be greatly magnified at the fork ends and will be responsible for poor handling characteristics. It may also cause the machine to fail in any machine examination test. It is important that the pivot pin is a good sliding fit, without any suspicion of play. Once again, these bushes are an extremely tight fit and should be taken to a Norton agent where they will be removed using a press.

G15CS MODELS
On the scrambles frame, the swinging arm consists of two 'Oilite' bearings in a steel sleeve. The steel sleeve is retained by one (two on earlier models) cotter pin. Remove the pin and press out the steel sleeve complete with the bearings.

12 Swinging arm rear suspension - reassembly

1   Reassemble the swinging arm fork by reversing the dismantling procedure. Metal-type bearings should be well lubricated with a heavy oil on assembly and at regular intervals. (eg. 140 EP). The 'Clayflex' type of bearing requires no maintenance. However a thin smear of grease on the metal surfaces will aid assembly. Do not lubricate as oil will tend to rot the rubber bushing.

13 Rear suspension units - examination

1   To remove or change the compression spring, it is necessary to grip the lower end of the unit in a vice fitted with soft clamps. Rotate the adjustable cam ring with a C or peg spanned until the unit is in the light load (solo riding) position. Then, with the help of a second person, compress the spring downwards so that the split collars can be displaced from the top of the shroud. If the spring pressure is released, the spring and shroud can be lifted away.

2   Reassemble the suspension unit by reversing the dismantling procedure.

3   If it is necessary to renew or change a spring for one of a different rating, it is important that both rear suspension units are treated alike. If the units are not well balanced, roadholding will be impaired.

14 Rear suspension units - adjusting the setting

1   Girling rear suspension units are fitted as standard. They have a three-position cam ring built into the lower portion of the leg, so that spring tension can be varied to suit existing load conditions. The lowest position should suit the average solo rider, under normal road conditions. When a pillion passenger is carried, the second or middle position offers a better choice and for continuous high speed work or competition events, the highest position is recommended.

The recommended settings are:-

Position 1 (least tension)	Normal solo riding
Position 2 (middle setting)	High speed touring
Position 3 (greatest tension)	High speed competition events or with pillion passenger and/or heavy loads.

2   The adjustments can be effected without need to detach the units. A C spanner or metal rod can be used to rotate the cam ring until the desired setting is obtained. It follows that the setting must be identical on both units, in the interests of good handling.

15 Engine Mountings

1   The engine and gearbox are mounted on steel engine plates. In the interests of lightness, Dural engine plates are marketed by private firms.

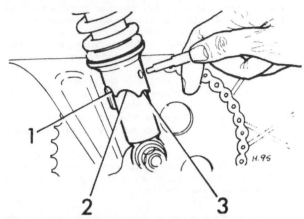

Fig. 6.4. Settings for rear suspension units

2   Maintenance of the engine mountings is required only to check that all the bolts are tight. (Torque settings 25 ft lb).

3   During overhauls, check that no cracking of the engine plates has occurred and look for obvious signs of ovality of all the fixing points. Oval attachments will tend to amplify the high natural vibration levels of a parallel twin engine.

16 Centre stand - Examination

1   The centre stand is mounted from lugs welded to the bottom frame loop. The pivot is two short nuts and bolts which abut the engine plates and are locked by two central through bolts. Check regularly for tightness.

2   The adjustments can be effected without need to detach the machine is pushed forward after parking, the centre stand will spring back into the fully retracted position and permit the machine to be wheeled, prior to riding.

3   The condition of the return spring and the return action should be checked frequently, also the security of the retaining nuts and bolts. If the stand drops whilst the machine is in motion, it may catch in some obstacle and unseat the rider.

17 Prop stand - examination

1   A prop stand which pivots from a lug bolted to the lower left hand frame tube provides an additional method of parking the machine when the centre stand is not used. This too has a return spring which should have sufficient tension to cause the stand to retract immediately the weight of the machine is lifted from it. It is important that this stand is examined at regular intervals, also the nut and bolt which act as the pivot. A falling prop stand can have serious consequences if it should fall whilst the machine is on the move.

2   The prop stand cannot be fitted after the engine plates.

18 Footrests - examination and renovation

1   The footrests, are malleable an will bend if the machine is dropped. Before thay can be straightened, they must be detached from the alloy plate and have the rubbers removed.

2   To straighten the footrests, clamp them in a vice and apply leverage from a long tube which slips over the end. The area in which the bend has occurred should be heated to a dull cherry red with a blow lamp during the bending operation; if the footrests are bent cold, there is risk of a sudden fracture.

19 Speedometer - removal and replacement

1   A Smiths speedometer is fitted to all models, calibrated in miles per hour or kilometres per hour (Continental models). An internal lamp is privided for illimination the dial and the odometer has a trip setting so that the lower of the two mileage recordings can be set to zero before a run is commenced.

2   The base of the speedometer has two studs, which permit it to be mounted via a 'U' bracket inside the headlamp shell. (This applies to models not supplied with tachometers - see Section 21).

3   Apart from defects in the drive or the drive cable itself, a speedometer which malfunctions is difficult to repair. Fit a replacement or alternatively entrust the repair to an instrument repair specialist, bearing in mind that the speedometer must function in a satisfactory manner to meet statutory requirements.

4   If the odometer readings continue to show an increase, without the speedometer indicating the road speed, it can be assumed the drive and drive cable are working correctly and that the speedometer head itself is at fault.

20 Speedometer cable - examination and renovation

1   It is advisable to detach the speedometer drive cable from time to time in order to check whether it is adequately lubricated, and whether the outer covering is compressed or damaged at any point along its run. A jerky or sluggish speedometer movement can often be attributed to a cable fault.

2   To grease the cable, withdraw the inner cable from the top. After removing the old grease, clean with a petrol soaked rag and examine the cable for broken strands or other damage.

3   Regrease the cable with high melting point grease and ensure that there is no grease on the last six inches, at the end where the cable enters the speedometer head, If this precaution is not observed, grease will work into the speedometer head and immobilise the movement.

4   Inspection will show whether the speedometer drive cable has broken. If so, the complete cable (inner and outer) must be renewed. Measure the cable length exactly when purchasing a replacement, because this measurement is critical.

20.1 Unscrew at rear wheel to remove speedometer cable. Note: grease nipple for periodic lubrication

21 Tachometer - removal and replacement

1   The tachometer drive is taken from the right hand end of the camshaft and emerges via a right angle gearbox mounted on the outside of the timing cover.

2   It is not possible to effect a satisfactory repair to a defective tachometer head, hence replacement is necessary if the existing head malfunctions. Make sure an exact replacement is obtained some tachometer heads work at half-speed if a different type of drive gearbox is employed.

3   The tachometer head is illuminated internally so that the dial can be read during the hours of darkness. It is mounted alongside the speedometer on either an aluminium plate which fits under the large bolts at the top of the forks or on a plate bolted to rubber mountings in the top fork yoke. The latter helps reduce vibration levels which may cause an inaccurate reading and shorten the life of components.

22 Tachometer drive cable - examination and renovation

Although a little shorter in length, the tachometer drive cable is identical in construction to that used for the speedometer drive. The advice given in Section 21 of this Chapter applies also to the tachometer drive cable.

23 Dualseat - removal

1   The seat is attached to the rear frame solely by a Dzus fastener through the rear of the seat and by two prongs which project from the crossmember of the frame behind the tank and locate in two rubber bushed lugs which project down from the front of the seat. Additional location is provided by lateral pegs which have rubber mats between the seat and the frame. The inside of the seat is foam rubber which will absorb water if the covering is torn or split.

2   Wideline dualseats are retained by two nuts from the underside, behind the tool tray.

24 Steering head lock

1   A steering head lock is fitted to post-1964 models. If the forks are turned fully to either the right or the left, they can be locked in that position as a precaution against theft.

2   Add an occasional few drops of thin machine oil to keep the lock in good working order. This should be added to the periphery of the moving drum and NOT the keyhole.

25 Cleaning - general

1   After removing all surface dirt with a rag or sponge which is washed frequently in clean water, the application of car polish or wax will restore a good finish to the cycle parts of the machine after they have dried thoroughly. The plated parts should require only a wipe with a damp rag, although it is permissible to use a chrome cleaner if the plated surfaces are badly tarnished.

2   Oil and grease, particularly when they are caked on, are best removed with a proprietary cleanser such as Gunk or Jizer. A few minutes should be allowed for the cleanser to penetrate the film of oil and grease before the parts concerned are hosed down. Take care to protect the magneto, carburettor(s) and electrical parts from the water, which may otherwise cause them to malfunction.

3   Polished aluminium alloy surfaces can be restored by the application of Solvol Autosol or some similar polishing compound, and the use of a clean duster to give the final polish.

4   If possible, the machine should be wiped over immediately after it has been used in the wet, so that it is not garaged under damp conditions which will promote rusting. Make sure to wipe the chain and if necessary re-oil it, to prevent water from enter-

ing the rollers and causing harshness with an accompanying high rate of wear. Remember there is little chance of water entering the control cables if they are lubricated regularly, as recommended in the Routine Maintenance section.

26 Carriers

Luggage racks and carriers are available with special mountings for the featherbed frame which is ideally designed to support rear loads. These items will transform a high-speed roadburner into an ideal touring machine in a few hours. Remember however, that a heavy rear-mounted load will affect machine stability to some extent.

27 Sidecar attachment

The featherbed frame has been extensively used for attaching sidecars and modified successfully to make racing outfits. However the standard of rigidity obtainable is not normally required and a lighter frame with a 600 cc engine was introduced in 1956 for 3 years, especially for sidecar work. If a featherbed frame model is required to tow a sidecar, Norton Villiers recommend some modifications to ensure the handling is compatible to the three wheel arrangement. Contact your Norton Villiers agent for details.

28 Fault diagnosis

Symptom	Cause	Remedy
Machine is unduly sensitive to road conditions	Forks and/or rear suspension units have defective damping	Check oil level in forks. Replace rear suspension units
Machine tends to roll at low speeds	Steering head bearings overtight or damaged	Slacken bearing adjustment. If no improvement, dismantle and inspect bearings
Machine tends to wander, steering is imprecise	Worn swinging arm bearing or excess clearances in engine mountings	Check and if necessary renew bearings
Fork action stiff	Fork legs have twisted in yokes or have been drawn together at lower ends	Slacken off spindle nut clamps, pinch bolts in fork yokes and fork top nuts. Pump forks several times before retightening from bottom. Is distance piece missing from fork spindle?
Forks judder when front brake is applied	Worn fork bushes Steering head bearings too slack	Strip forks and replace bushes Re-adjust, to take up play
Wheels out of alignment	Frame distorted as result of accident damage	Check frame alignment after stripping out If bent, specialist repair is necessary

Chapter 7 Wheels, brakes and tyres

Contents

Specifications

Wheel sizes

Front (Except G15 Mk II)	19 in. dia. WM2-19 rim
Front (G15 Mk II)	18 in. dia. WM3-18 rim
Rear (Except those below)	19 in. dia. WM2-19 rim
Rear (Atlas, G15 Mk II, G15CS,	
N15CS and some early 650 std models)	18 in. dia. WM3-18 rim

Tyre sizes

Front (19 in. rim)	3.00 x 19 in. (Except Atlas)
	3.25 x 19 in. (Atlas)
Front (18 in. rim)	3.25 x 18 in.
Rear (19 in. rim)	3.50 x 19 in.
Rear (18 in. rim)	4.00 x 18 in.

Tyre pressures

Front	25 psi* (1.76 kg/cm^2)
Rear	22 psi* (1.55 kg/cm^2)

Note: All 650 cc 750 cc Featherbed framed models and 'SS' models should be equipped with an Avon GP tyre pr equivalent (ie Dunlop K81, TT100). Tyre pressure should be 24 psi normally and increased to 30 psi for sustained speeds in excess of 110 mph.

*Pressure will vary according to cross section of tyre and model to which the tyre is fitted. The figures given are an approximate guide.

Brakes

Front	8 x 1¼ ins. internal diameter*
	(single leading shoe)
Rear	7 x 1¼ ins. internal diameter
	(single leading shoe)

*A twin leading shoe brake plate is available as an optional extra.

† Wheel bearings

Front, left side bearing	(Hoffman 117 DR)
	17 x 40 x 12 mm.
Front, right side bearing	(Hoffman 117 DR)
	17 x 40 x 12 mm
Rear, left side bearing	(Hoffman 117 DR)
	17 x 40 x 16 mm
Rear, right side bearing	(Hoffman 177)
	17 x 40 x 12

Rear chain

General standard:

Early models (pre 1966)	97 links, 5/8 x 1/4 in.
Later models (post 1966)	97 links, 5/8 x 3/8 in.
G15CSR	98 links, 5/8 x 3/8 in.

† except P11 models

1 General description

All Featherbed framed models are normally equipped with 19 in diameter wheels front and rear. The "650 American" and "750 Scrambles" machines have an 18 in rear wheel (see specifications) the wheel hubs being common to all machines. The rear wheel is of the quickly-detachable type and can be removed from the frame without need to disturb the final drive sprocket and brake drum.

In general, a 3.00ins. section front tyre is adequate for touring work on all machines except the 750s where a 3.25 ins tyre is recommended. This tyre should also be used where high speed road work and heavy braking is anticipated. The rear tyre should be either a 3.50 x 19 in or 4.00 x 18 in depending on the model (and the rim).

All machines are fitted with internally expanding drum brakes. The rear drum is a 7in diameter single leading shoe unit. The front brake is normally a similar unit of 8 in diameter; however both the factory and many private firms offer a twin leading shoe unit for higher braking efficiency.

2 Front wheel - examination and renovation

1 Place the machine on the centre stand so that the front wheel is raised clear of the ground. Spin the wheel and check for rim alignment. Small irregularities can be corrected by tightening the

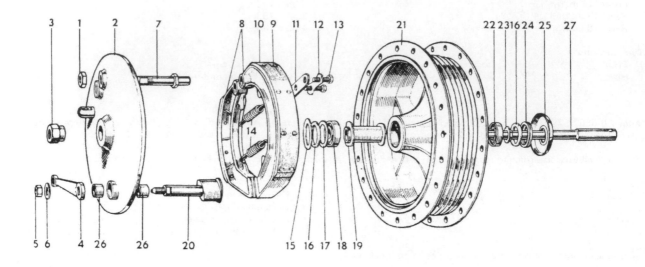

FIG. 7.1 FRONT WHEEL ASSEMBLY

1 Nut for torque stop pivot pin
2 Front tube brake plate
3 Nut for hub spindle
4 Front brake cam lever
5 Nut for cam lever
6 Washer for lever
7 Torque stop pivot pin for brake plate
8 Hub brake shoe and lining 2 off
9 Brake lining rivet - 16 off

10 Hub brake shoe lining - 1 pair
11 Brake shoe pivot pin retaining plate
12 Tag washer for retaining plate
13 Bolt for retaining plate 2 off
14 Brake shoe return spring 2 off
15 Steel plate for hub bearing felt washer

16 Hub bearing felt washer 2 off
17 Hub bearing pen steel washer
18 Hub bearing, right hand
19 Bearing distance tube
20 Front brake cam
21 Front hub shell and brake drum

22 Hub bearing, left hand
23 Hub bearing distance piece (plain side)
24 Hub bearing lock ring
25 Hub bearing dust cap
26 Brake cam bush 2 off
27 Front wheel spindle

spokes in the affected area, although a certain amount of experience is necessary if over-correction is to be avoided. Any 'flats' in the wheel rim should be evident at the same time. These are more difficult to remove with any success and in most cases the wheel will have to be rebuilt on a new rim. Apart from the effect on stability, especially at high speeds, there is much greater risk of damage to the tyre beads and walls if the machine is ridden with a deformed wheel.

2   Check for loose or broken spokes. Tapping the spokes is the best guide to tension. A loose spoke will produce a quite different note and should be tightened by turning the nipple (right hand thread). Always recheck for run-out by spinning the wheel again.

---

3   Front drum brake assembly - examination, renovation and reassembly (Single Leading shoe)

1   The front wheel can be removed from the forks and the brake assembly complete with brake plate detached by following the procedure given in Chapter 6, Section 2, when the machine is raised on a wooden box.

2   Before dismantling the brake assembly, examine the condition of the brake linings. If they are wearing thin or unevenly, they must be replaced. The angle of the brake operating arm will

give the best indication of the degree of wear, if checked when the brake is applied. ie. If the brake unit has been correctly assembled, the angle between the brake operating arm and the brake cable should not exceed 90°when the brake is applied If the angle does exceed 90 °, dismantle and check the linings.

3   To remove the brake shoes from the brake plate, first detach the retaining plate which bolts to the brake shoe pivot or torque stop pins. They are secured with a tab washer. Lever off the brake shoe return springs with a screwdriver and separate the brake shoes.

4   It is possible to replace the brake linings fitted with rivets and not bonded on, as is the current practice. Much will depend on the availability of the original type of linings; service-exchange brake shoes with bonded-on linings may be the only practical form of replacement.

5   If new linings are fitted by rivetting, it is important that the rivet heads are countersunk, otherwise they will rub on the brake drum and be dangerous. Make sure the lining is in very close contact with the brake shoe during the rivetting operation; a small C clamp of the type used by carpenters can often be used to good effect until all the rivets are in position. Finish off by chamfering off the end of each shoe, otherwise fierce brake grab may occur due to the pick-up of the leading edge of each lining.

6   Before replacing the brake shoes, check that the brake operating cam is working smoothly and not binding in its pivot. The cam can be removed for greasing by unscrewing the nut on the

end of the operating arm, after marking the arm and the brake plate so that they are replaced in an identical position. Draw the operating arm off the squared end of the operating cam. The operating cam will withdraw from the inside of the brake plate, permitting the shaft and the bush in which the cam operates to be cleaned and greased.

7   Check the inner surface of the brake drum, on which the brake shoes bear. The surface should be smooth and free from indentations, or reduced braking efficiency is inevitable. Remove all traces of brake lining dust and wipe the surface with a rag soaked in petrol to remove any traces of grease or oil. DO NOT USE PARAFFIN.

8   To reassemble the brake shoes on the brake plate, fit the shoes on the brake plate in their correct positions and lever the return springs back into position, making sure they are correctly located. Do not omit to replace the retaining plate and tab washer.

4   Front brake assembly - examination, renovation and re-assembly (Twin leading shoe)

1   The maintenance of this unit may be carried out with the aid of the previous section, a little common sense and the following notes.

2   When withdrawing the operating arms from the squared ends of the operating cams (ref. 3.6), do not slacken or remove the rod which joins the arms or else the relationship between the two arms will be lost and will have to be reset on assembly.

3   To reset the rod length such that both brake shoes operate together, in order to achieve maximum braking effect, refer to Section 10.2 of this Chapter.

5   Front wheel bearings - examination and renovation

1   When the brake plate assembly has been withdrawn, unscrew the locking ring which retains the left hand bearing. The locking ring has a right hand thread and can be unscrewed by using Norton Villiers service tool 063965 - a peg spanner - or by the use of a pin punch and hammer. If the ring proves difficult to remove, warm the hub and apply the peg spanner or punch in different holes.

2   After the locking ring has been unscrewed, lift out the felt seal and distance piece. Then insert the front wheel spindle from the right hand (brake drum) side of the hub to drive the right hand double row bearing inwards whilst at the same time displacing the left hand single row bearing from the hub. A rawhide mallet should be used to drive the spindle so that the displaced bearing only just clears the hub. If this precaution is not observed, there is risk of damaging the spacer.

3   Working from the other side of the wheel, enter the wheel spindle through the spacer tube which is still within the hub and drive the double row bearing outwards again, whilst holding the whole assembly in a central position. The bearing will be displaced from the hub, together with the felt retaining washer, felt seal, outer washer and spacer.

4   Remove all the old grease from the hub and bearings and give the latter a final wash in petrol. Check the bearings for play or signs of roughness when they are turned. If there is any doubt about their condition, play safe and renew them. A new bearing has no discernible play.

5   Before replacing the bearings, first pack the hub with new, high melting point grease and the bearings themselves. Locate the single row bearing first in the left hand side of the hub, then fit the distance washer (plain side to bearing) the felt seal and the locking ring. Tighten the ring fully with either the peg spanner or the pin punch.

6   Insert the bearing spacer tube from the right, small end first, so that it presses against the bearing already positioned. Pass the

3.2 The front brake plate lifts out

3.3 The brake shoes are retained by a retaining plate, tab washer and two bolts

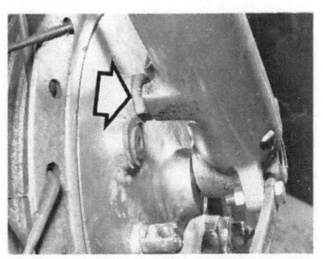
3.8 When replacing the front wheel, the brake stop must align as shown

5.1 Unscrew the left hand locking ring

5.3a Right hand bearing felt seal is retained by two large washers

5.3b Drive out bearing (double row) ...

5.3c ... and remove the spacer

front wheel spindle through the double row bearing and enter this bearing into the right hand end of the hub. Push the wheel spindle further into the hub so that it passes through the spacer tube and through the centre of the left hand bearing, until the head abuts the end of the double row bearing. Drive the end of the spindle so that the double row bearing enters the hub fully and abuts the spacer tube. Refit the smaller felt retaining washer, the felt seal and the large steel washer. This latter washer is either peened or pushed into position, depending on the type fitted.

7   Do not use excessive force to displace the wheel bearings. If there is the remotest risk of damaging the wheel spindle, use instead a drift of the same diameter.

6   Rear wheel - examination, removal and renovation

1   Before removing the rear wheel, check for rim alignment, damage to the rim and loose or broken spokes by following the procedure described in Section 2 of this Chapter.

2   To remove the rear wheel without disturbing the final drive chain, place the machine on the centre stand so that the rear wheel is clear of the ground. Disconnect the speedometer drive cable from the gearbox through which the rear wheel spindle passes, then unscrew the spindle from the right hand side of the

machine (right hand thread). When the spindle is withdrawn, the spacer will fall free and the speedometer gearbox can be pulled off the end of the rear wheel hub.

3   Before the wheel spindle is slackened and withdrawn, it is necessary to remove the three rubber blanking plugs from the right hand side of the hub and the three sleeve nuts contained within the tunnels blanked off by the plugs. The wheel can then be pulled off the studs which project from the rear of the brake drum, after the rear wheel spindle is slackened and withdrawn from the right.

4   Occasions arise when it is necessary to remove the rear wheel complete with brake drum and sprocket, in which case a different procedure is necessary. Commence by supporting the machine on the centre stand, so that the rear wheel is clear of the ground. Then disconnect the rear chain at the split link, a task made easier if the link is positioned in the teeth of the rear wheel sprocket.

5   Disconnect the rear brake cable rod by removing the adjuster nut and pulling the rod through the trunnion in the end of the brake operating arm. Disconnect the speedometer drive cable, from the gearbox through which the rear wheel spindle passes, then slacken and withdraw the wheel spindle from the right hand side of the machine. The spacer will fall free when the spindle is withdrawn and the speedometer gearbox can be pulled off the

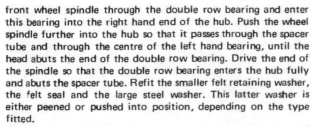

end of the hub.

6 Remove the chainguard. Unscrew and remove the left hand spindle nut and pull the wheel over to the right hand side of the swinging arm fork so that the torque stop of the brake will disengage from the slotted lug of the left hand fork tube. The wheel is now free to be lifted from the rear of the machine. Note:- In both cases, removal of the rear wheel is eased if the machine is raised so that the wheel may be withdrawn without fouling the mudguard.

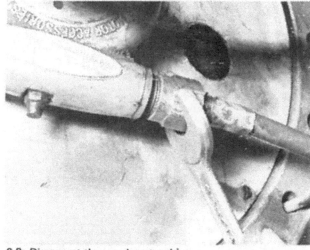

6.2a Disconnect the speedometer drive

6.2b Unscrew rear wheel spindle from right hand side. Note: spacer

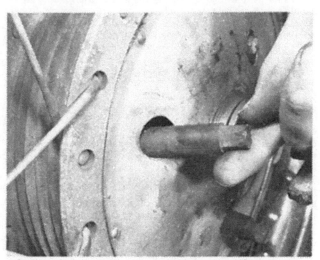

6.3a Remove rubber blanking plugs and unscrew sleeve nuts

6.3b Pull wheel off the three locating studs

6.4 Machine must be supported high enough to allow wheel to be withdrawn from under mudguard

7 Rear wheel - dismantling, examining and reassembling the hub

1 Before the hub can be dismantled further, it is first necessary to separate the brake assembly from the brake drum and then the brake drum from the wheel hub.

2 Remove the three blanking plugs from the right hand end of the hub and unscrew the three sleeve nuts contained within the tunnels blanked off by the plugs. The studs which project from the back of the brake drum are then freed, permitting the brake drum to be lifted away.

3 To dismantle the rear hub, unscrew the bearing locking ring found on the right hand (speedometer drive) end of the hub. The ring has a LEFT HAND thread, and it is necessary to use either Norton Villiers service tool 063965 or a pin punch and hammer. When the ring has unscrewed, lift out the distance piece and felt seal.

4 Replace the thick washer on the rear wheel spindle and the right hand spacer which fits between the speedometer gearbox and the fork end. Insert the spindle through the bearing in the left hand (brake drum) end of the hub, with the washer and spacer against the spindle head, then drive the bearing into the hub as far as it will go by hitting the end of the spindle with a rawhide mallet. This will commence the displacement of the right hand bearing from the hub. Withdraw the rear wheel spindle and its attachments, then insert the front wheel spindle in its place, threaded end foremost, from the same side. If the spindle is held horizontally and the end tapped with the mallet, the right hand bearing and bearing spacer will be displaced completely from the hub.

5 Insert the rear wheel spindle and thick washer into the wheel hub from the opposite (right hand) side and drive the left hand bearing outwards from the hub, complete with the felt retaining washer, felt seal and dished washer.

6 Remove all old grease from the hub and bearings, then check the bearings for signs of play or roughness as described in Section 5.4. Replace the bearings.

7 To reassemble the hub, fit the right hand bearing first so that the felt seal and locking ring can be fitted and the latter tightened fully. Special care is necessary to ensure the slots which transmit the drive to the speedometer gearbox are not damaged. Then insert the spacer tube into the hub from the left, noting that the ends are of different lengths and that the longer end will locate with the bearing on the right hand side of the hub. Insert the left hand bearing and drive it into the hub, applying loud only to the outer race. Fit the felt retaining washer, felt seal and dished washer, which is either peened or pushed into position, depending on the type fitted.

8 Rear brake assembly - examination, renovation and reassembly

1 The rear brake assembly, which is of the drum type can be withdrawn complete with brake plate, after the rear wheel has been withdrawn from the frame. It is similar in construction to the front wheel drum brake.

2 Follow an identical procedure for dismantling, examining and reassembling the brake to that described in Section 3 of this Chapter, with the following notes.

3 A large spacer exists between the brake plate and the frame fork. This spacer surrounds the dummy spindle which is a clearance fit in the drum.

4 When replacing the brake drum and plate on the fork end, ensure that the brake stop engages with the slot on the inside of the left hand fork leg.

5 Note that the rear wheel sprocket is an integral part of the brake drum. If the sprocket teeth are worn, chipped, broken or hooked, the brake drum must be renewed, preferably in conjunction with a new chain and gearbox final drive sprocket.

7.1 Bearing locking ring has left hand thread

7.4 Drive out the right hand bearing and spacer complete

7.5a Drive out felt retaining washers ...

7.5b ... followed by the left hand double row bearing

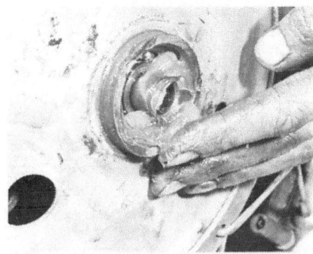

7.6 Repacking the bearings with grease

8.3a Ease off the brake plate distance piece (may be a tight fit)

8.3b Spindle and brake plate pull out from the brake drum

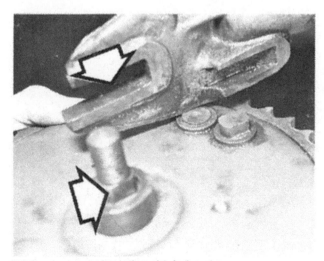

8.4 Ensure spindle flats align with fork ends

9 Rear wheel - replacement

1 The rear wheel is replaced in the frame by reversing the dismantling procedure described in Section 6 of this Chapter.

2 Make sure the distance piece is fitted between the hub and the right hand side of the swinging arm fork and that the speedometer drive gearbox has located correctly with the drive slots in the hub centre, before the wheel spindle is inserted. Check that the brake torque stop has engaged with the lug on the left hand fork tube.

3 If the brake drum has been removed from the rear wheel, make sure the three sleeve nuts are tightened fully after it has been replaced. If the nuts work loose, the studs with thich they engage will be subjected to a shear stress. It is advisable to check the tightness of these nuts periodiaclly, even if the wheel has not been removed from the frame.

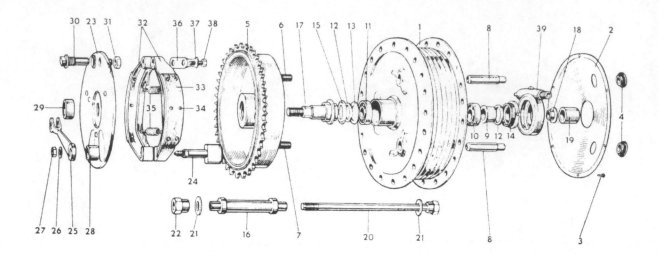

FIG. 7.2 REAR WHEEL ASSEMBLY

1 Rear hub only
2 Diaphragm for rear hub
3 Drive screws for diaphragm - 6 off
4 Rubber grommets for rear hub - 3 off
5 Rear brake drum with studs (43 teeth)
6 Locating stud for brake drum
7 Non-locating studs for brake drum - 2 off
8 Sleeve nut for brake drum - 3 off

9 Hub bearing distance piece
10 Hub bearing, right hand
11 Hub bearing, left hand
12 Felt washer for bearing 2 off
13 Pen steel washer for bearing
14 Hub bearing lock ring
15 Hub bearing felt washer retaining washer
16 Rear hub inner sleeve
17 Brake drum attachment piece
18 Speedometer drive distance piece

19 Hub spindle distance piece
20 Rear hub spindle
21 Washer for hub spindle 2 off
22 Nut for hub spindle
23 Rear brake plate
24 Rear brake cam
25 "   "   " lever
26 Washer for lever
27 Nut for lever
28 Rear brake cam bearing
29 Rear brake plate distance piece
30 Rear brake torque pivot pin

31 Nut for torque pivot pin
32 Hub brake shoe with lining (rear) - 2 off
33 Hub brake shoe lining (rear) - 1 pair
34 Brake lining rivets (14 per set) - 1 set
35 Brake shoe return spring - 2 off
36 Brake shoe pivot pin retaining plate
37 Tab washer for retaining plate
38 Bolt for retaining plate 2 off
39 Speedometer drive gearbox

10 Adjusting the front brake

1   All models fitted with drum front brakes are cable operated. A cable adjuster is provided on the brake plate. Adjustment is a matter of individual requirements but it should not be possible for the end of the brake lever to touch the handlebars before the brake is applied fully, or even approach very close. On the other hand and brake shoes should not rub on the brake drum when the brake is not in use. Apart from causing reduced performance, the brake shoes may overheat and give rise to brake fade, rendering the brake useless in an emergency. Turn either adjuster outwards (anticlockwise) to take up wear which would otherwise cause increasing slackness in the brake cable.

2   Twin leading shoe brakes only

The screwed operating rod which joins the two brake operating arms of the front brake should not require attention unless the setting has been disturbed. It is imperative that the leading edge of each brake shoe comes into contact with the brake drum simultaneously, if maximum braking efficiency is to be achieved. Check by detaching the clevis pin from the eye of one end of the threaded rod, hold both levers on with two

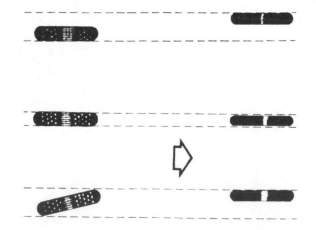

Fig. 7.3. Checking wheel alignment

spanners and turn the adjuster on the threaded rod until the clevis pin can just be reinserted.

Replace the clevis pin and do not omit the split pin through the end which retains it in position. Re-check the brake lever adjustment before the machine is tried on the road.

3 Check that the brake pulls off correctly when the handlebar lever is released. Sluggish action is usually due to a poorly lubricated brake cable, broken return springs or a tendency for the brake operating cams to bind in their bushes. Dragging brakes affect engine performance and can cause severe overheating of both the brake shoes and wheel bearings. A badly adjusted twin leading shoe brake has less braking effect than a single leading shoe brake.

11 Adjusting the rear brake

1 The rear brake is of the single leading shoe type and has a single adjuster at the end of the brake operating arm. Turn the adjuster nut inwards (clockwise) to decrease the brake pedal movement and take up excess slack in the rod. The amount of free movement in the brake pedal is a matter of personal preference, but it should not be necessary to depress the pedal far before the brake is applied fully. Beware of adjusting the brake too closly so that the linings are still in rubbing contact with the brake drum.

2 After adjusting the rear brake, it is sometimes necessary to adjust the stop lamp switch, so that the stop lamp lights up at the correct time. The switch is bolted to a lug on the frame which carries the pillion footrest.

12 Final drive chain - examination, lubrication and adjustment

1 Except on a few models, the final drive chain does not have the benefit of full enclosure or positive lubrication which is afforded to the primary drive chain. In consequence, it will require attention from time to time, particularly when the machine is used on wet or dirty roads.

2 Chain adjustment is correct when there is approximately ¾ inch play in the middle of the run. Always check at the tightest spot on the chain run, under load.

3 If the chain is too slack, adjustment is effected by slackening the rear wheel spindle and nut, then pushing the wheel back-

11.1 The rear brake rod adjuster

11.2 Plate on brake rod operates rear brake light

wards by means of the chain adjusters at the end of the rear fork. Make sure each adjuster is turned an equal amount, so that the rear wheel is kept centrally-disposed within the frame. When the correct adjusting point has been reached, push the wheel hard forward to take up any slack, then tighten the spindle and nut. Re-check the chain tension and the wheel alignment, before the final tightening of the spindle and nut. (Fig. 7.3).

4 Application of engine oil from time to time will serve as a satisfactory form of lubrication, but it is advisable to remove the chain every 500 miles (unless it is enclosed within a chaincase, in which case every 2000 miles should suffice) and clean it in a bath of paraffin before immersing it in a special chain lubricant such as Linklyfe or Chainguard. These latter types of lubricant achieve better and more lasting penetration of the chain links and rollers and are less likely to be thrown off when the chain is in motion.

5 To check whether the chain is due for replacement, lay it lengthwise in a straight line and compress it, so that all play is taken up. Anchor one end and then pull on the other, to stretch the chain in the opposite direction. If the chain extends by more that 1/8 inch per foot, renewal is necessary.

6 When replacing the chain, make sure the spring link is positioned correctly, with the closed end facing the direction of travel, and the convex side out. Reconnection is made easier if the ends of the chain are pressed into the rear wheel sprocket.

7 Note: As with the primary chain, it is more preferable to run with the chain slack than too tight.

8 Remember that any adjustment of the primary drive will effect the tension of the final drive chain.

13 Wheel balance

1 On any high performance machine it is important that the front wheel is balanced, to offset the weight of the tyre valve. If this precaution is not observed, the out-of-balance wheel will produce an unpleasant hammering that is felt through the handlebars at speeds from approximately 50 mph upwards.

2 To balance the front wheel, place the machine on the centre stand so that the front wheel is well clear of the ground and check that it will revolve quite freely, without the brake shoes rubbing. In the unbalanced state, it will be found that the wheel always comes to rest in the same position, with the tyre valve in the six o'clock position. Add balance weights to the spokes diametrically opposite the tyre valve until the tyre valve is counterbalanced, then spin the wheel to check that it will come to rest in a random position on each occasion. Add or subtract

weight until perfect balance is achieved.

3   Only the front wheel requires attention. There is little point in balancing the rear wheel because it will have little noticeable effect on the handling of the machine.

4   Balance weights of various sizes which will fit around the spoke nipples are available from Norton Motors. If difficulty is experienced in obtaining them lead wire or even strip solder can be used as an alternative, kept in place with insulating tape.

14 Tyres - removal and replacement

1   At some time or other the need will arise to remove and replace the tyres, either as the result of a puncture or because replacements are necesary to offset wear. To the inexperienced, tyre changing represents a formidable task yet if a few simple rules are observed and the technique learned, the whole operation is surprisingly simple.

2   To remove the tyre from either wheel, first detach the wheel from the machine by following the procedure in Chapters 6.2 or depending on whether the front or the rear wheel is involved. Deflate the tyre by removing the valve insert and when it is fully deflated, push the bead from the tyre away from the wheel rim on both sides so that the bead enters the centre well of the rim. Remove the locking cap and push the tyre valve into the tyre itself.

3   Insert a tyre lever close to the valve and lever the edge of the tyre over the outside of the wheel rim. Very little force should be necessary; if resistance is encountered it is probably due to the fact that the tyre beads have not entered the well of the wheel rim all the way round the tyre.

4   Once the tyre has been edged over the wheel rim, it is easy to work around the wheel rim so that the tyre is completely free on one side. At this stage, the inner tube can be removed.

5   Working from the other side of the wheel, ease the other edge of the tyre over the outside of the wheel rim which is furthest away . Continue to work around the rim until the tyre is free completely from the rim.

6   If a puncture has necessitated the removal of the tyre, re-inflate the inner tube and immerse it in a bowl of water to trace the source of the leak. Mark its position and deflate the tube. Dry the tube and clean the area around the puncture with a petrol soaked rag. When the surface has dried, apply rubber solution and allow this to dry before removing the backing from the patch and applying the patch to the surface.

7   It is best to use a patch of the self-vulcanising type, which will form a very permanent repair. Note that it may be necessary to remove a protective covering from the top surface of the patch, after it has sealed in position. Inner tubes made from synthetic rubber may require a special type of patch and adhesive, if a satisfactory bond is to be achieved.

8   Before replacing the tyre, check the inside to make sure the agent which caused the puncture is not trapped. Check the outside of the tyre, particularly the tread area, to make sure nothing is trapped which may cause a further puncture.

9   If the inner tube has been patched on a number of past occasions, or if there is a tear or large hole, it is preferable to discard it and fit a replacement. Sudden deflation may cause an accident, particularly if it occurs with the front wheel.

10   To replace the tyre, inflate the inner tube sufficiently for it to assume a circular shape but only just. Then push it into the tyre so that it is enclosed completely. Lay the tyre on the wheel at an angle and insert the valve through the rim tape and the hole in the wheel rim. Attach the locking cap on the first few threads,

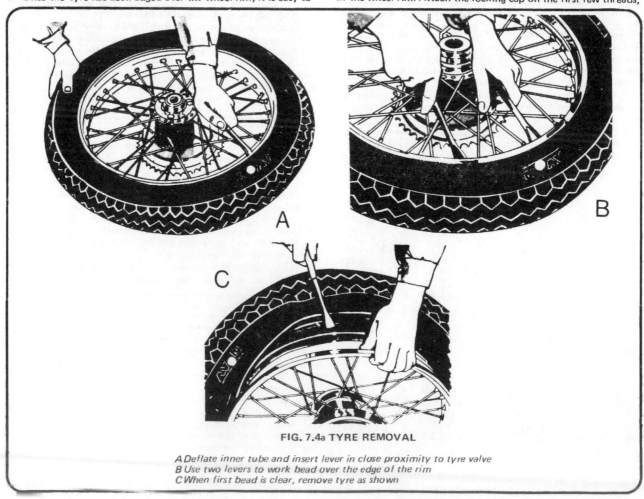

FIG. 7.4a TYRE REMOVAL

A Deflate inner tube and insert lever in close proximity to tyre valve
B Use two levers to work bead over the edge of the rim
C When first bead is clear, remove tyre as shown

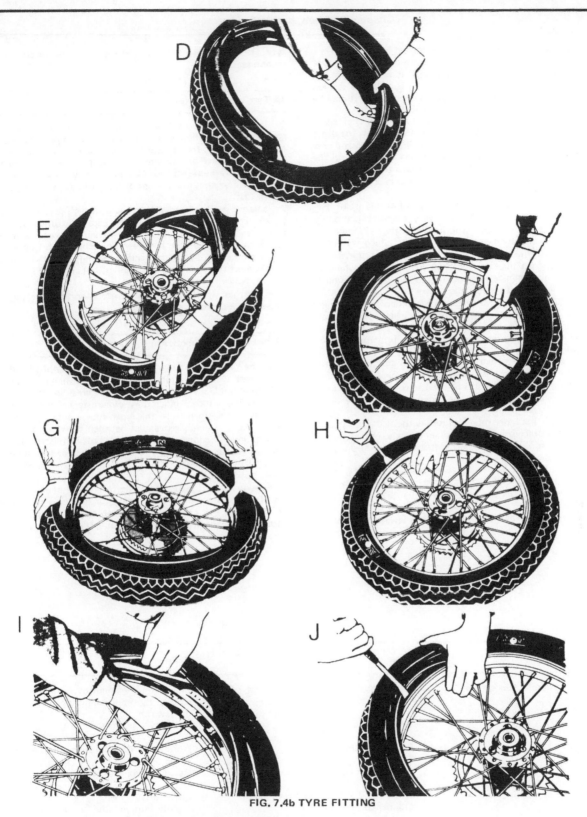

FIG. 7.4b TYRE FITTING

D   Inflate inner tube and insert in tyre
E   Lay tyre on rim and feed valve through hole in rim
F   Work first bead over rim, using lever in final section
G   Use similar technique for second bead. Finish at tyre valve position
H   Push valve and tube up into tyre when fitting final section, to avoid trapping

Security bolts
I   Fit the security bolt very loosely when one bead of the tyre is fitted
J   Then fit tyre in normal way. Tighten bolt when tyre is properly seated

sufficient to hold the valve captive in its correct location.

11 Starting at the point furthest from the valve, push the tyre bead over the edge of the wheel rim until it is located in the central well. Continue to work around the tyre in this fashion until the whole of one side of the tyre is on the rim. It may be necessary to use a tyre lever during the final stages.

12 Make sure there is no pull on the tyre valve and again commencing with the area furthest from the valve, ease the other bead of the tyre over the edge of the rim. Finish with the area close to the valve, pushing the valve up into the tyre until the locking cap touches the rim. This will ensure the inner tube is not trapped when the last section of the bead is edged over the rim with a tyre lever.

13 Check that the inner tube is not trapped at any point. Re-inflate the inner tube, and check that the tyre is seating correctly around the wheel rim. There should be a thin rib moulded around the wall of the tyre on both sides, which should be equidistant from the wheel rim at all points. If the tyre is unevenly located on the rim, try bouncing the wheel when the tyre is at the recommended pressure. It is probable that one of the beads has not pulled clear of the centre well.

14 Always run the tyres at the recommended pressures and never under or over-inflate. The correct pressures for solo use are given in the Specifications Section of this Chapter. If a pillion passenger is carried, increase the rear tyre pressure as recommended.

15 Tyre replacement is aided by dusting the side walls particularly in the vicinity of the beads, with a liberal coating of French chalk. Washing-up liquid can also be used to good effect, but this has the disadvantage of causing the inner surfaces of the wheel rim to rust.

16 Never replace the inner tube and tyre without the rim tape in position. If this precaution is overlooked thate is good chance of the ends of the spoke nipples chafing the inner tube and causing a crop of punctures.

17 Never fit a tyre which has a damaged tread or side walls. Apart from the legal aspects, there is a very great risk of a blow-out, which can have serious consequences on any two-wheel vehicle.

18 Tyre valves rarely give trouble, but it is always advisable to check whether the valve itself is leaking before removing the tyre. Do not forget to fit the dust cap which forms an effective second seal.

15 Tyre valve dust caps

1 Tyre valve dust caps are often left off when a tyre has been replaced, despite the fact that they serve an important two-fold function. Firstly, they prevent dirt or other foreign matter from entering the valve and causing the valve to stick open when the tyre pump is next applied. Secondly, they form an effective second seal so that in the event of the tyre valve sticking, air will not be lost.

2 Isolated cases of sudden deflation at high speeds have been traced to the omission of the dust cap. Centrifugal force has tended to lift the tyre valve off its seating and because the dust cap is missing and there had been no second seal. Racing inner tubes contain provision for this happening because the valve inserts are fitted with stronger springs, but standard inner tubes do not, hence the need for the dust cap.

3 Note that when a dust cap is fitted for the first time, the wheel may have to be rebalanced.

16 Security bolt - function and fitting

1 If the drive from a high-powered engine is applied suddenly to the rear wheel of a motor cycle, wheel spin will occur with and initial tendency for the wheel rim to creep in relation to the tyre and inner tube. Under these circumstances there is risk of the valve being torn from the inner tube, causing the tyre to deflate rapidly, unless movement between the rim and tyre can be restrained in some way. A security bolt fulfills this role in a simple and effective manner, by clamping the bead of the tyre to the well of the wheel rim so that any such movement is no longer possible.

2 A security bolt is fitted to the rear wheel of many Norton models. Before attempting to remove or replace a tyre, it must be slackened off completely so that the clamping action is released. The accompanying tyre fitting illustrations show how the security bolt is fitted and secured.

17 Fault diagnosis

Symptom	Cause	Remedy
Handlebars oscillate at low speeds	Buckle or flat in wheel rim, most probably front wheel.	Check rim alignment by spinning wheel. Correct by retensioning spokes or rebuilding on new rim.
	Tyre not straight on rim	Check tyre alignment.
Machine lacks power and accelerates	Brakes binding	Worn brake drum provided best evidence. Re-adjust brakes.
	Chains too tight	Re-adjust chains
Brakes grab when applied gently	Ends of brake shoes not chamfered	Chamfer with file.
	Elliptical brake drum	Lightly skim in lathe (specialist attention required).
Brake pull-off sluggish	Brake cam binding in housing	Free and grease.
	Weak brake shoe springs	Renew if springs have not become displaced.
Harsh transmission	Worn or badly adjusted final drive chain	Adjust or renew as necessary
	Hooked or badly worn sprockets	Renew as a pair
	Loose rear wheel	Check wheel retaining bolts.

Chapter 8 Electrical system

Contents

Specifications

Battery

Voltage	6V
Manufacturer	Lucas
Type	Lead/Acid
Model	*MLZ9E
Capacity	12 ampere/hr
Earth connection	Positive

* Lucas PUZ5A used on P11, P11A models only
  Lucas PUZ7E - early machines, standard models
    Exide 3EV9 -   "   "   De Luxe models

Alternator

Manufacturer	Lucas	
Type	RM 15 and RM 19	

Rectifier

Manufacturer	Lucas	
Type	47132 and 49072	

Zener Diode (12v systems only)

Manufacturer	Lucas	

Horn

Manufacturer	Lucas	
Type	HF 1441	

Stop/Tail lamp

Manufacturer	Lucas	
Type	Mod. 564	

Bulbs

	6V	12V
Voltage	6V	12V
Manufacturer	Lucas	Lucas
Main Headlamp		
Home	30/24W	45/40W
French Export	36/36W	
Continental	35/35W	
Other Export	30/24W	
Pilot	3W	6W
Stop/Tail Lamp	6/18W	6/21W
Speedometer/Tachometer	2.0W	2.2W (L643)

1 General description

Four types of electrical system have been fitted to the Norton twins since their inception is 1949. The original system, which was used until 1957, had a dynamo and a voltage regulator to provide and regulate the current. These early models are becoming increasingly rare, and it is recommended that in the event of electrical trouble, the owner should check the wiring system against the diagram included and the condition of the carbon brushes in the dynamo; if they both appear acceptable, take the machine to a reputable auto-electrical dealer, to have the system checked. Since 1958, all models have been fitted with either a 6 volt (pre 1964) or 12 volt system (most post-1964 models) comprising a crankshaft driven alternator with a six pole permanent magnet rotor which rotates within a six pole laminated iron stator coil assembly, a rectifier to convert the alternating current into direct current for charging the battery, and the battery itself. On 12 volt models, voltage regulation is achieved by means of a Zener diode, a semi conductor device which becomes conductive in the reverse flow direction when a predetermined voltage level is reached.

2 Crankshaft alternator- checking the output

1 An ammeter mounted in the top of the headlamp shell of all models provides visual indication of the output from the alternator, whilst the engine is running. If no charge is indicated, even with a full lighting load, the performance of the alternator is suspect and should be investigated by disconnecting the alternator leads at the connector and putting a bulb between the white/green lead and earth, the green/yellow lead and earth, and if of the three wire type the green/black wire and earth. If the bulb lights in each of the two (or three) positions when the engine is running, the alternator is providing output.

2 As a first precaution, note the colour coding and disconnect the three leads from the alternator; clean the "bullet" connections and the snap connector internals in order to remove all oil and water deposits. Replace the connections and retest.

3 Using a simple battery and bulb type tester, check using the wiring as per the wiring diagram provided (after DIS-CONNECTING THE BATTERY).

4 If the tests in 2 & 3 do not isolate the electrical trouble,

remove the primary chaincase for a visual examination of the alternator. Check that

1) The grommet in the inner chaincase is in good condition and the output leads are not shorting to the chaincase or the frame.

2) Broken chain rollers have not been impelled into the stator windings.

If everything appears in order and the battery connections, the rectifier, the Zener Diode (12 volt models only) and the fuse (most recent models only) all appear satisfactory, take the machine to a reputable auto-electrical dealer or Norton specialist and have the alternator output checked. The equipment required for these tests is not normally available to "home mechanics" and, hence, the test procedure is not covered here.

5 Alternatively fit a substitute alternator and re-check. If the alternator proves to be defective, a new replacement can be obtained at reduced cost, through the Lucas Service Exchange Scheme. If the alternator still apparently gives no output it may be the ignition/lighting switch at fault. This is another item which can be speedily checked by an expert with the correct equipment.

3 Battery - examination and maintenance

1 Early models have a single 6 volt battery, normally of 12 amp/t.our capacity. The later 12 volt systems employ two 6 volt batteries in series. If in doubt about the battery used on the various models, consult either a reputable auto-electrical dealer or a Norton—Villiers agent. It is important to obtain the correct capacity battery in order to poivide good lighting and maximum life from the battery. Note:- all batteries are POSITIVE EARTH.

2 Battery maintenance is limited to keeping the electrolyte level just above the plates and separators, as denoted by the level line marked on the case. The level of the electrolyte is readily visible, due to the translucent nature of the case. Do not overfill and make sure the vent pipe is attached, (if fitted) so that it will discharge away from any parts of the machine liable to suffer damage from corrosion.

3 Unless acid is split, which may occur if the machine falls over, use only distilled water for topping up purposes, until the correct level is restored. If acid is spilt on any part of the machine, it should be beutralised immediately with an alkali such as washing soda or baking powder, and washed away with plenty of water. This will prevent corrosion from taking place. Top up in this instance with sulphuric acid of the correct specific gravity (1.260 - 1.280).

4 It is seldom practicable to repair a cracked battery case because the acid which is already seeping through the crack will prevent the formation of an effective seal, no matter what sealing compound is used. It is always best to replace a cracked battery, especially in view of the risk of corrosion from the acid leakage.

5 Make sure the battery is clamped securely A loose battery will vibrate and its working life will be greatly shortened, due to the paste being shaken out of the plates.

6 Ensure the battery connections are kept clean and tight. If corrosion occurs, clean the parts concerned by immersing them in a solution of baking powder to neutralise the acid. When reconnection is made, apply a light smearing of petroleum jelly such as vaseline, to prevent corrosion recurring.

7 If the battery should become discharged, a temporary boost (on 6 volt models) can be effected during daylight running by connecting the alternator output leads as follows:-

a) Reconnect the "green and yellow" and the "green and black" connectors

b) Reconnect the "green and black" to the "green and yellow" leads

c) DO NOT interfere with the "green and white".

This is a temporary measure and will damage the battery if persisted with; if the battery will not hold the "boost" charge, it probably requires renewal. This condition is usually indicated by grey, sulphated plates and loose deposits in the bottom of the battery.

4 Silicon rectifier - function and testing

1 The function of the silicon rectifier is to convert the AC current from the alternator into DC current for charging the battery. The rectifier is of the full wave type and does not require any maintenance.

2 The rectifier is located close to the battery, in the compartment below the dualseat. It is located in this position so that it is not directly exposed to water or to accidental damage. One of the most frequent causes of rectifier failure is caused by the inadvertent connection of the battery in reverse, which initiates a reverse flow of current through the rectifier. It is not practicable to repair a damaged rectifier; renewal is the only satisfactory solution.

3 The usual indication of rectifier failure is the inability to keep the battery charged and poor lighting in general. To check whether the rectifier is functioning correctly, take the machine to a reputable auto-electrical dealer or a Norton Villiers agent.

4 If for any reason the rectifier has to be removed, care should be taken to ensure the plates are not twisted in relation to one another, or even scratched or bent. It is only too easy to damage the internal connections and render the rectifier unfit for further service. The centre bolt must ALWAYS be held firmly with a spanner, whilst the mounting nut is removed or replaced. ie. The torque on the rectifier plates is preset by the manufacturer and should not be disturbed for any reason.

5 Fuse - location and replacement late models only

1 A 35 amp fuse is fitted into the battery negative lead, close to the battery carrier. It is retained within a nylon holder and is of the replaceable type. To release the fuse, push both halves of the fuse holder body together and twist.

2 If a fuse blows, it should be replaced, after checking to ensure that no obvious short circuit has occurred. If the second fuse blows shortly afterwards the electrical circuit must be checked thoroughly, to trace and eliminate the fault.

3 Always carry at least one spare fuse. If a situation arises where a fuse blows whilst the machine is in use and no spare is available, a 'get you home' remedy is to remove the defective fuse and wrap it in silver paper before replacing it in the fuse holder. The silver paper will restore electrical continuity by bridging the broken fuse wire. This expedient should NEVER be used if there is evidence of a short circuit or other major electrical fault, otherwise more serious damage will be caused. Replace the blown fuse at the earliest possible opportunity, to restore full circuit protection.

4 For similar reasons, do not fit a fuse which has a higher rating than the original, or it may no longer form the weakest link in the circuit protection chain.

6 Zener diode - function, location and testing

12 volt models only.

1 The Zener diode performs the voltage regulating function accepting excess current from the alternator which is not required for battery charging purposes and converting it into heat which is dissipated by the mounting (heat sink) to which the diode is attached. It follows, therefore, that the Zener diode must be located so that it is rigidly attached to a good heat conductive base which will dissipate the heat by its location in a continuous air stream. It is clamped to a bracket from the front left-hand tank fixing bolt. If repositioning or including a Zener diode in the circuit, ensure that it is mounted such that it does not restrict the movement of the handlebars nor the travel of the forks. It must also be in the airstream, to assist cooling.

2 Contact between the base of the Zener diode and the heat sink must be clean and corrosion-free, otherwise the heat path will be interrupted. The tightening torque is critical. It must not exceed 2 ft. lb. or there is risk of the mounting stud shearing.

3 To check the operation of the diode, connect a dc ammeter between the snap connector and feed cable, positive lead to the diode terminal and negative lead to the feed cable. Connect a dc voltmeter between the diode terminal and earth, negative lead to the terminal and the positive lead to earth. Ensure the lights are switched off and that there is no other electrical load applied, then start the engine and gradually increase the speed. The Zener diode must be replaced if the ammeter shows a reading before the voltmeter records 12.75 volts, or if the voltmeter shows more that 15.5 volts before the ammeter reading had reached 2 amps. If the Zener diode is faulty, it must be renewed. A repair is not possible.

4 Throughout the test, the battery must be in a fully charged condition, or a completely false reading may result. If the state of charge is low, fit a fully-charged battery for the duration of the test.

7 Headlamp - renewing bulbs and adjusting beam height

1 The headlamp is fitted with a sealed beam unit in which the main bulb is of the pre-focus type. The complete unit is secured to the headlamp rim by wire clips; access is gained by slackening the screw in the top of the headlamp shell (close to the rim) and then lifting the rim away with the sealed beam unit attached. A push-fit connector forms the detachable electrical connections with the main headlamp bulb, located within the sealed unit. If either the main or dipped filaments fail, the main bulb must be renewed. The bulb holder is released by pressing inwards and turning to the left. Note: it will only engage in one position.

2 The pilot bulb and bulb holder are attached to the reflector of the sealed beam unit. The pilot bulb has a bayonet fitting and is easily renewed when the bulb holder is pulled out of the reflector.

3 Beam height is adjusted by slackening the two bolts which retain the headlamp shell to the fork shrouds and tilting the headlamp either upwards or downwards. Adjustments should always be made with the rider seated normally.

4 UK lighting regulations stipulate that the lighting system must be arranged so that the light does not dazzle a person standing in the same horizontal plane as the vehicle, at a distance greater than 25 yards from the lamp, whose eye level is not less that 3 feet 6 inches above that plane. It is easy to approximate this setting by placing the machine 25 yards away from a wall, on a level road, and setting the beam height so that it is concentrated at the same height as the distance from the centre of the headlamp to the ground. The rider must be seated normally during this operation, and the pillion passenger, if one is carried regularly.

5 If the sealed beam unit is broken, it can be removed from the headlamp rim by detaching the wire retaining clips, after the rim has been removed from the front of the headlamp.

6 A pilot bulb of either 12 volt 6 watt or 6 volt 3 watt rating is fitted, depending on the model. The fitting is the miniature bayonet cap type; ensure that the rubber sealing washer is replaced between the bulb holder and the light reflector when reassembling.

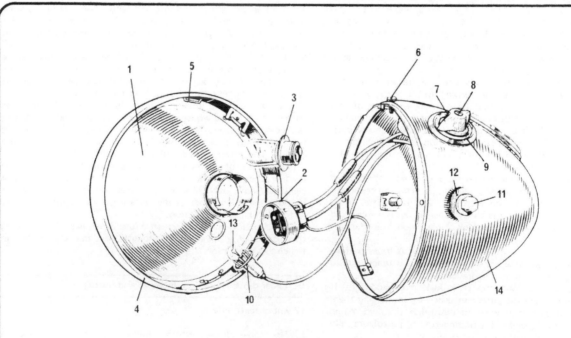

FIG. 8.1 HEADLAMP ASSEMBLY

1 Reflector/glass unit	5 Wire spring for retaining	8 Screw for item 7	12 Washer for item 11 - 2 off
2 Main bulbholder	items 1 and 4 - 6 off	9 Bezel plate	13 Pilot bulb
3 Main bulb	6 Rim locating screw	10 Pilot bulbholder	14 Headlamp shell
4 Rim	7 Lighting switch knob	11 Shell fixing/adjusting	
		screw - 2 off	

7.1a Slacken centre/top screw to release headlamp unit

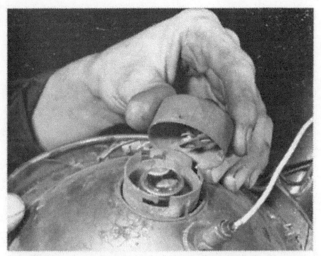

7.1b Push in and turn anticlockwise to release main bulb holder

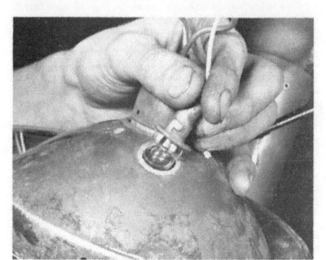

7.2 Pilot bulb holder is push fit in reflector

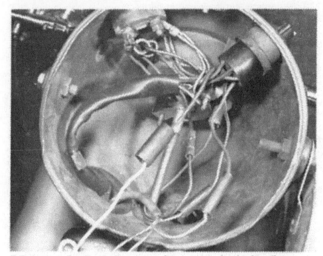

7.3 Beam adjusting bolts mate with set nuts in the headlamp shell

8   Lighting and Ignition Switch

1   The lighting switch is carried on the left hand side of the headlamp shell or on the speedometer/rev. counter bracket. On coil ignition models, the lighting and ignition switches are combined; this unit (Lucas PR58) is extremely complicated and should not be dismantled.

2   The lighting switch on magneto and dynamo models may be dismantled for cleaning. Remove the screw which retains the position knob; pull off the knob complete with washers. The nut which retains the switch body to the headlamp shell is now exposed; undo the nut and push the switch into shell.

3   Screw connections fix the leads to the switch. Before dismantling, check the wiring against the wiring diagram and ensure the numbers and colours are readily distinguishable.

4   On coil ignition models, always ensure the ignition is switched off before leaving the machine or the battery will discharge itself.

9   Tail and stop lamps - renewing bulbs

1   Removal of the plastic lens cover, which is retained by two barrel nuts, will reveal the bulb holder which holds the combined tail and stop lamp. The bulb is of the bayonet fitting type, but has offset pins to prevent accidental inversion. The tail lamp filament is rated at 6 watts and the stop lamp filament at 18 watts (21 watts on 12 volt models).

2   The stop lamp switch actuated by the rear brake is attached to the brake pedal. It is adjusted as described in Chapter 7, Section 11.2.

10   Dipswitch

1   The dipswitch is operated by the left hand thumb and mounted on the handlebar by means of a clip secured by a nut and bolt.

2   Faulty operation is normally due to either dirty contacts or a broken cam or cam-springs within the switch if the unit malfunctions it must be renewed. It is seldom possible to effect a satisfactory repair.

3   Always ensure the electrical connections are tight and always replace the rubber insulating mat between the switch and the handlebar; the latter also helps keep oil and water out of the switch.

4   After servicing a switch, a small amount of grease on the moving parts will aid operation. Do not allow the grease to contaminate the contacts.

9.1a Two barrel nuts retain the rear reflector lens

9.1b Rear bulb has offset pins

11 Speedometer and tachometer bulbs

The speedometer and tachometer heads each have an internal bulb to illuminate the dial during the hours of darkness. The bulb holder is a puch ift in the underside of each instrument. A bulb of the miniature bayonet fitting type, is fitted to each.

12 Horn adjustment

1   The horn is suspended from either the cylinder head steady or from the top of the front downtubes of the frame.
2   If the horn performance is sub-standard check the following points:-
a)   The state of charge of the battery
b)   The wiring system for loose or dirty connections and short circuits
c)   The horn push for loose connections, dirty contacts or a bad earth.

3   Horn adjustment is dependent upon the type of horn fitted. It is recommended that the adjustment is carried out at a Lucas Service Station since incorrect adjustment may place a very heavy load on the battery.

13 Wiring - layout and examination

1   The wiring is colour-coded and will correspond with the accompanying wiring diagrams.
2   Visual inspection will show whether any breaks or frayed outer coverings are giving rise to short circuits. Another source of trouble may be the snap connectors, particularly where the connector has not been pushed home fully in the outer casing. Early models are especially prone to wiring faults because the rubber-covered cables used at that period deteriorate as time progresses.
3   Intermittent short circuits can sometimes be traced to a chafed wire which passes through a frame member. Avoid tight bends in the wire or situations where the wire can become trapped or stretched between casings.

14 Fault diagnosis

Symptom	Cause	Remedy
Complete electrical failure	Blown fuse Isolated battery	Check wiring and electrical components for short circuit before fitting new 15 amp fuse. Check battery connections, also whether connections show signs of corrosion
Dim lights, horn inoperative	Discharged battery	Recharge battery with battery charger and check whether alternator is giving correct output
Constantly 'blowing' bulbs	Vibration, poor earth connection	Check whether bulb holders are secured correctly. Check earth return or connections to frame

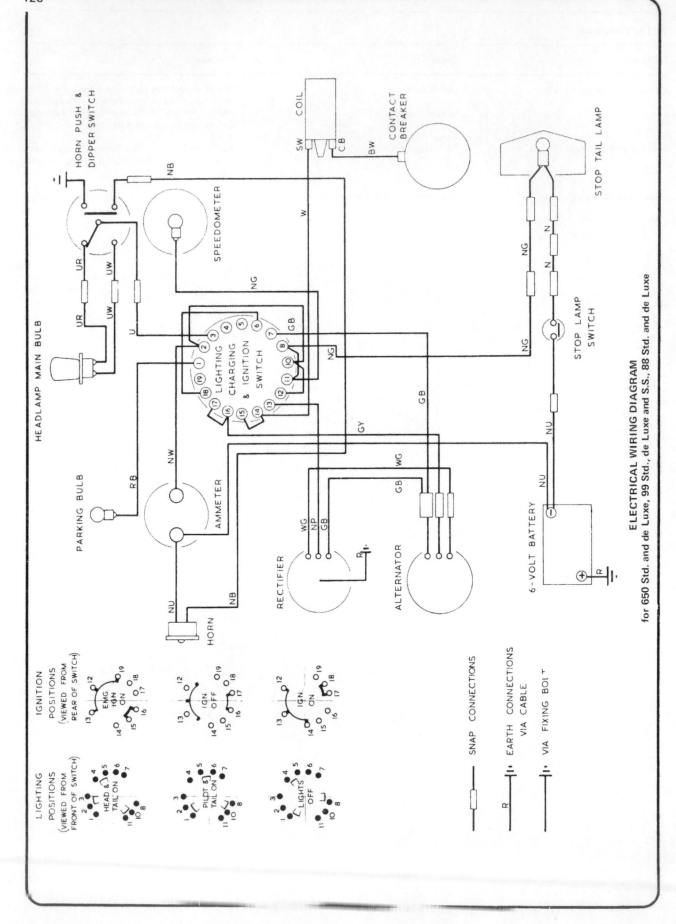

ELECTRICAL WIRING DIAGRAM
for 650 Std. and de Luxe, 99 Std., de Luxe and S.S., 88 Std. and de Luxe

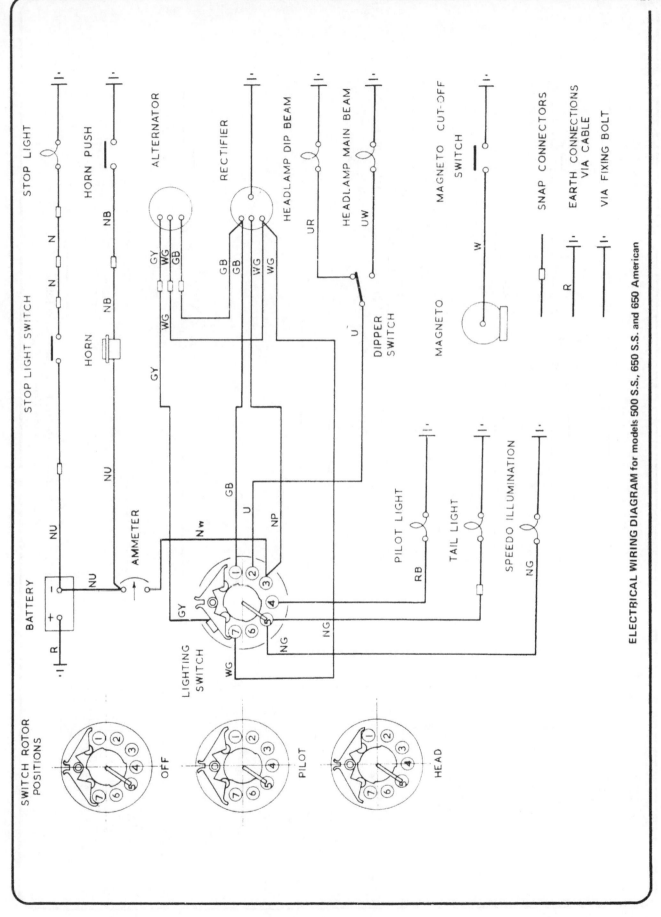

ELECTRICAL WIRING DIAGRAM for models 500 S.S., 650 S.S. and 650 American

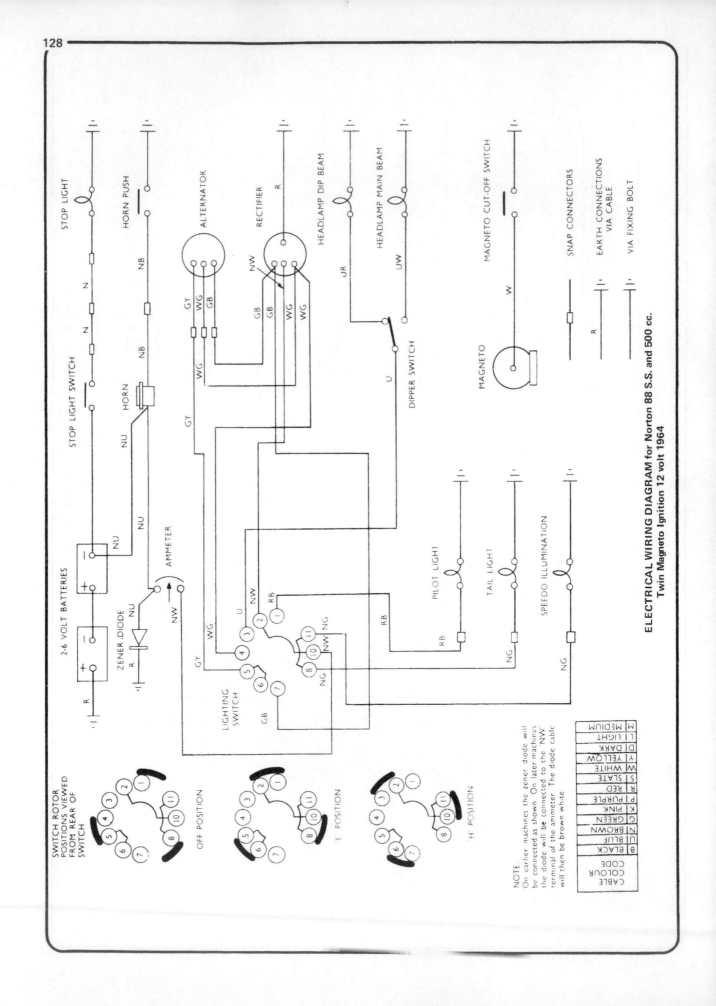

ELECTRICAL WIRING DIAGRAM for Norton 88 S.S. and 500 cc.
Twin Magneto Ignition 12 volt 1964

129

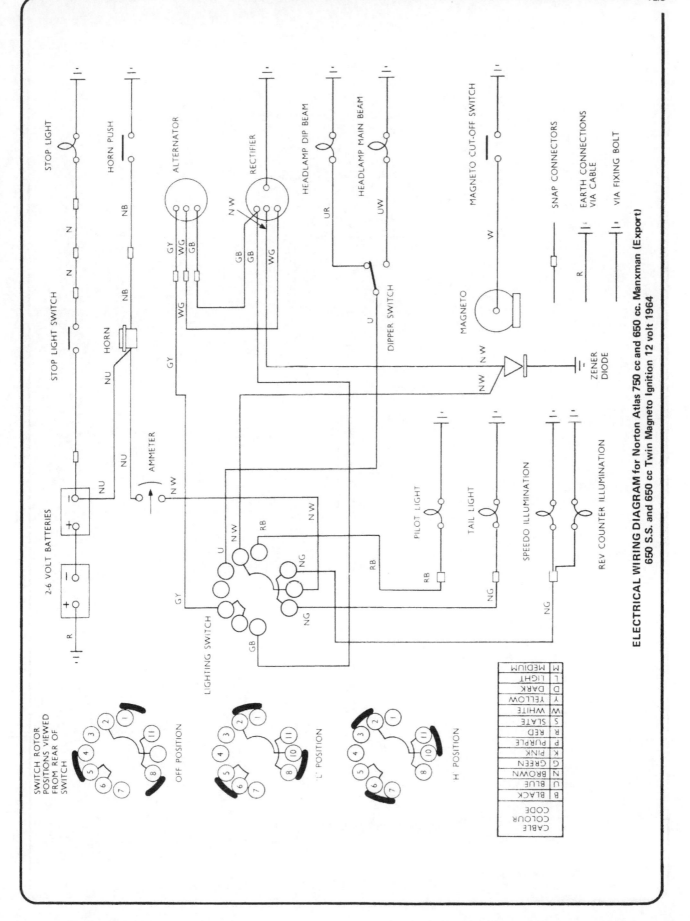

ELECTRICAL WIRING DIAGRAM for Norton Atlas 750 cc and 650 cc. Manxman (Export) 650 S.S. and 650 cc Twin Magneto Ignition 12 volt 1964

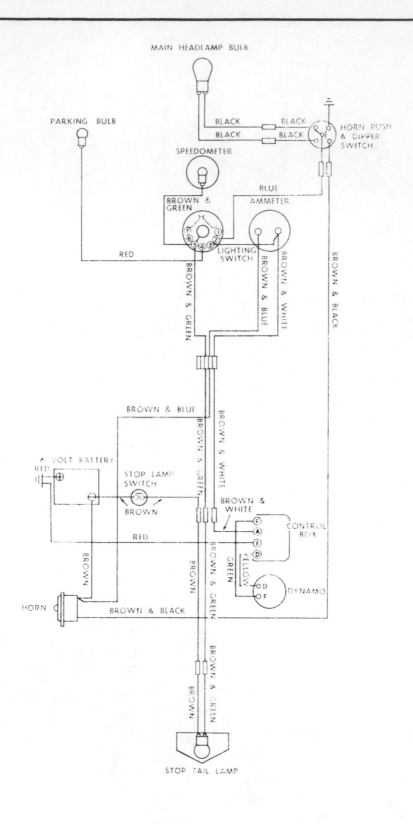

1956 - 1957 MODELS 88 and 99

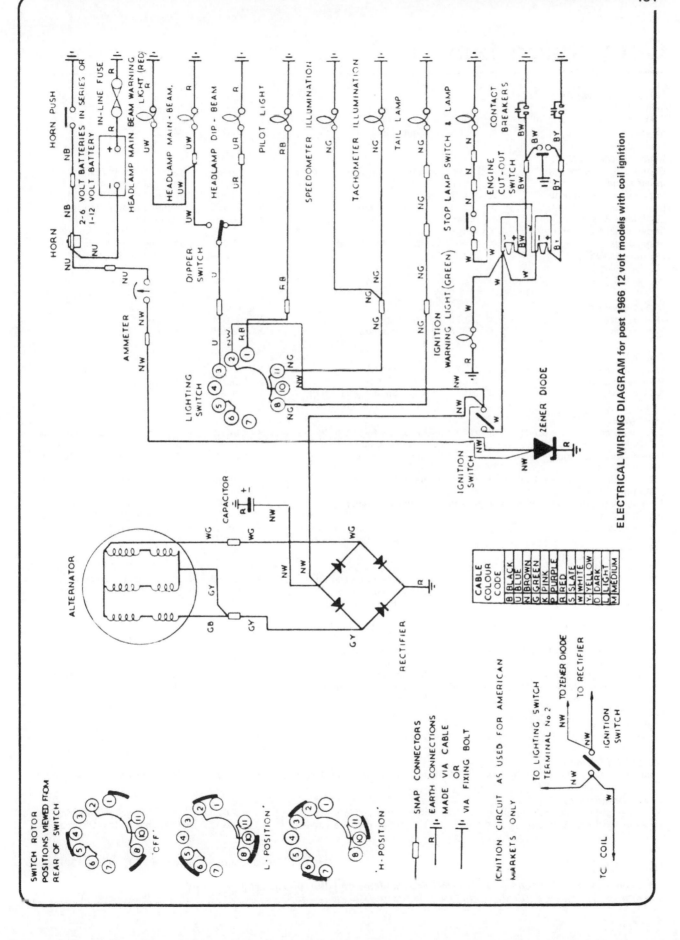

ELECTRICAL WIRING DIAGRAM for post 1966 12 volt models with coil ignition

Conversion factors

Length (distance)
Inches (in)	X 25.4	= Millimetres (mm)	X 0.0394	= Inches (in)
Feet (ft)	X 0.305	= Metres (m)	X 3.281	= Feet (ft)
Miles	X 1.609	= Kilometres (km)	X 0.621	= Miles

Volume (capacity)
Cubic inches (cu in; in^3)	X 16.387	= Cubic centimetres (cc; cm^3)	X 0.061	= Cubic inches (cu in; in^3)
Imperial pints (Imp pt)	X 0.568	= Litres (l)	X 1.76	= Imperial pints (Imp pt)
Imperial quarts (Imp qt)	X 1.137	= Litres (l)	X 0.88	= Imperial quarts (Imp qt)
Imperial quarts (Imp qt)	X 1.201	= US quarts (US qt)	X 0.833	= Imperial quarts (Imp qt)
US quarts (US qt)	X 0.946	= Litres (l)	X 1.057	= US quarts (US qt)
Imperial gallons (Imp gal)	X 4.546	= Litres (l)	X 0.22	= Imperial gallons (Imp gal)
Imperial gallons (Imp gal)	X 1.201	= US gallons (US gal)	X 0.833	= Imperial gallons (Imp gal)
US gallons (US gal)	X 3.785	= Litres (l)	X 0.264	= US gallons (US gal)

Mass (weight)
Ounces (oz)	X 28.35	= Grams (g)	X 0.035	= Ounces (oz)
Pounds (lb)	X 0.454	= Kilograms (kg)	X 2.205	= Pounds (lb)

Force
Ounces-force (ozf; oz)	X 0.278	= Newtons (N)	X 3.6	= Ounces-force (ozf; oz)
Pounds-force (lbf; lb)	X 4.448	= Newtons (N)	X 0.225	= Pounds-force (lbf; lb)
Newtons (N)	X 0.1	= Kilograms-force (kgf; kg)	X 9.81	= Newtons (N)

Pressure
Pounds-force per square inch (psi; lbf/in^2; lb/in^2)	X 0.070	= Kilograms-force per square centimetre (kgf/cm^2; kg/cm^2)	X 14.223	= Pounds-force per square inch (psi; lbf/in^2; lb/in^2)
Pounds-force per square inch (psi; lbf/in^2; lb/in^2)	X 0.068	= Atmospheres (atm)	X 14.696	= Pounds-force per square inch (psi; lbf/in^2; lb/in^2)
Pounds-force per square inch (psi; lbf/in^2; lb/in^2)	X 0.069	= Bars	X 14.5	= Pounds-force per square inch (psi; lbf/in^2; lb/in^2)
Pounds-force per square inch (psi; lbf/in^2; lb/in^2)	X 6.895	= Kilopascals (kPa)	X 0.145	= Pounds-force per square inch (psi; lbf/in^2; lb/in^2)
Kilopascals (kPa)	X 0.01	= Kilograms-force per square centimetre (kgf/cm^2; kg/cm^2)	X 98.1	= Kilopascals (kPa)

Torque (moment of force)
Pounds-force inches (lbf in; lb in)	X 1.152	= Kilograms-force centimetre (kgf cm; kg cm)	X 0.868	= Pounds-force inches (lbf in; lb in)
Pounds-force inches (lbf in; lb in)	X 0.113	= Newton metres (Nm)	X 8.85	= Pounds-force inches (lbf in; lb in)
Pounds-force inches (lbf in; lb in)	X 0.083	= Pounds-force feet (lbf ft; lb ft)	X 12	= Pounds-force inches (lbf in; lb in)
Pounds-force feet (lbf ft; lb ft)	X 0.138	= Kilograms-force metres (kgf m; kg m)	X 7.233	= Pounds-force feet (lbf ft; lb ft)
Pounds-force feet (lbf ft; lb ft)	X 1.356	= Newton metres (Nm)	X 0.738	= Pounds-force feet (lbf ft; lb ft)
Newton metres (Nm)	X 0.102	= Kilograms-force metres (kgf m; kg m)	X 9.804	= Newton metres (Nm)

Power
Horsepower (hp)	X 745.7	= Watts (W)	X 0.0013	= Horsepower (hp)

Velocity (speed)
Miles per hour (miles/hr; mph)	X 1.609	= Kilometres per hour (km/hr; kph)	X 0.621	= Miles per hour (miles/hr; mph)

Fuel consumption*
Miles per gallon, Imperial (mpg)	X 0.354	= Kilometres per litre (km/l)	X 2.825	= Miles per gallon, Imperial (mpg)
Miles per gallon, US (mpg)	X 0.425	= Kilometres per litre (km/l)	X 2.352	= Miles per gallon, US (mpg)

Temperature
Degrees Fahrenheit = (°C x 1.8) + 32   Degrees Celsius (Degrees Centigrade; °C) = (°F - 32) x 0.56

*It is common practice to convert from miles per gallon (mpg) to litres/100 kilometres (l/100km), where mpg (Imperial) x l/100 km = 282 and mpg (US) x l/100 km = 235

English/American terminology

Because this book has been written in England, British English component names, phrases and spellings have been used throughout. American English usage is quite often different and whereas normally no confusion should occur, a list of equivalent terminology is given below.

English	American	English	American
Air filter	Air cleaner	Number plate	License plate
Alignment (headlamp)	Aim	Output or layshaft	Countershaft
Allen screw/key	Socket screw/wrench	Panniers	Side cases
Anticlockwise	Counterclockwise	Paraffin	Kerosene
Bottom/top gear	Low/high gear	Petrol	Gasoline
Bottom/top yoke	Bottom/top triple clamp	Petrol/fuel tank	Gas tank
Bush	Bushing	Pinking	Pinging
Carburettor	Carburetor	Rear suspension unit	Rear shock absorber
Catch	Latch	Rocker cover	Valve cover
Circlip	Snap ring	Selector	Shifter
Clutch drum	Clutch housing	Self-locking pliers	Vise-grips
Dip switch	Dimmer switch	Side or parking lamp	Parking or auxiliary light
Disulphide	Disulfide	Side or prop stand	Kick stand
Dynamo	DC generator	Silencer	Muffler
Earth	Ground	Spanner	Wrench
End float	End play	Split pin	Cotter pin
Engineer's blue	Machinist's dye	Stanchion	Tube
Exhaust pipe	Header	Sulphuric	Sulfuric
Fault diagnosis	Trouble shooting	Sump	Oil pan
Float chamber	Float bowl	Swinging arm	Swingarm
Footrest	Footpeg	Tab washer	Lock washer
Fuel/petrol tap	Petcock	Top box	Trunk
Gaiter	Boot	Torch	Flashlight
Gearbox	Transmission	Two/four stroke	Two/four cycle
Gearchange	Shift	Tyre	Tire
Gudgeon pin	Wrist/piston pin	Valve collar	Valve retainer
Indicator	Turn signal	Valve collets	Valve cotters
Inlet	Intake	Vice	Vise
Input shaft or mainshaft	Mainshaft	Wheel spindle	Axle
Kickstart	Kickstarter	White spirit	Stoddard solvent
Lower leg	Slider	Windscreen	Windshield
Mudguard	Fender		

Index